The Global Warming and Climate Change Superscam

Dr. Vincent Gray

The Global Warming Scam and Climate Change Superscam

© 2015 Vincent Gray All Rights Reserved
Print ISBN (grayscale) 978-1-941071-23-6
Print ISBN (color) 978-1-941071-24-3
ebook ISBN 978-1-941071-26-7

Other books by Vincent Gray
The Global Warming Delusion: A Critique of Climate Change 2001
Confessions of a Climate Sceptic

This book is sold subject to the condition that it shall not, by way of trade or otherwise, be lent, resold, hired out or otherwise circulated without the publisher's prior consent in any form of binding or cover other than that in which it is published and without a similar condition including this condition being imposed on the subsequent purchaser.

STAIRWAY PRESS—SEATTLE

Cover Design by Guy D. Corp
www.GrafixCorp.com

STAIRWAY PRESS

www.StairwayPress.com
1500A East College Way #554
Mount Vernon, WA 98273 USA

Dedicated to the memory of Augie Auer, Ernst Georg Beck, John Daly, Zbigniew Jaworowski, Bob Kay, Owen McShane and Peter Toynbee…

…who did not live to be told that they were right.

Acknowledgements

Parts of this book have appeared on the following websites:
New Zealand Climate Science Coalition
http://www.climatescience.org.nz/
Australian Climate Sceptics
http://theclimatescepticsparty.blogspot.co.nz/
Watts up with that http://wattsupwiththat.com/
Principia Scientific International http://www.principia-scientific.org

THE 5th HORSEMAN OF THE APOCALYPSE...
...CLIMATE CHANGE!
FOSSIL FUELS
TOM SCOTT 5/4/14

Contents

Introduction: The Gods Will Destroy	5
CHAPTER 1: THE CLIMATE	12
Weather Forecasting	13
Climate as a Heat Engine	18
Climate Properties	20
Solar Irradiance	20
The Lapse Rate	27
The Circulation System	31
Ocean Oscillations	33
Meteorology Today	36
Climatology	37
CHAPTER 2: ENVIRONMENTAL RELIGION	38
The Origins of Religion	38
Evolution Science	43
The Scientific Method	48
Induction	49
Deduction	52
Validation	55
Falsifiability	55
Opinions of Experts	56
Mathematical Models	56
Accuracy	57
Climate Science	58
Climatology	58
Climate Change Pseudoscience	59
CHAPTER 3: ENVIRONMENTAL SCAMS	62
Nuclear Winter	62
Silent Spring	64
World Dynamics and Limits to Growth	68
The Population Bomb	70
Depletion of Resources	72
Anti Fracking	75
Genetic Engineering	76
Acid Rain	78
Water Shortage	79

The Ozone Hole 81
Global Warming Scam and Climate Change Superscam 86
CHAPTER 4: THE IPCC SUPERSCAM 87
The Framework Convention on Climate Change 91
The Climate Change Superscam 92
Climate Change 94
Climate 96
Climate System 97
The IPCC Reports 99
Climate Change: The IPCC Scientific Assessment 1990 .. 101
The 1992 Supplementary Report 105
Climate Change 1994 108
Climate Change 1995: The Science of Climate Change 110
Summary 115
Section 8.1 115
Section 8 116
Section 8.2.2 Inadequate Representation of Feedbacks 117
Section 8.2.5 117
Section 8.3.2 117
Section 8.3.3.3 118
Section 8.4.1 118
Section 8.4.1.1 118
Section 8.4.1.3 119
Section 8.4.2.1 120
Section 8.4.2.3. 120
Section 8.5.2 121
Section 8.6 121
Section 8.7 123
The Special Report on Emissions Scenarios 2000 123
Climate Change 2001: The Scientific Basis 125
Climate Change 2007: The Physical Science Basis 128
Climate Change 2013: The Physical Science Basis 132
Controlling the Scientists 140
CHAPTER 5: THE GREENHOUSE EFFECT 146

Jean Baptiste Joseph Fourier 146
Horace-Bénédict de Saussure 148
Claude Pouillet 153
John Tyndall 154
Svante Arrhenius 159
Samuel Pierpont Langley 160
Guy Stewart Callendar 165
Sir George Simpson 167
Thomas Chrowder Chamberlin 168
Hubert Lamb 169
Gilbert Norman Plass 170
Roger Revelle 171
Robert W. Wood 172
The Real Greenhouse 175
Summary 176
CHAPTER 6: THE MODELS SCAM 177
  THE IPCC MODELS 181
  Radiative Forcing 188
  Evaluation of Models 192
  The IPCC Scenarios 195
  Climate Sensitivity 197
CHAPTER 7: THE GLOBAL WARMING SCAM 207
  Temperature 207
  Hansen's Initiative 217
  Sea Surface Temperature 223
  Microwave Sensing Units 227
  Fiddling the Figures 231
  United States Temperatures 241
  Quality of U.S. Stations 244
  The Hockey Stick 245
  The Latest IPCC Report 250
  AIRS Satellite Temperatures 253
  NASA's AQUA Platform 253

Conclusions 253
CHAPTER 8: CARBON DIOXIDE 255
　Atmospheric Carbon Dioxide Measurement 260
　Measurement of Downward Intensity 269
　Conclusions 270
CHAPTER 9: CLIMATEGATE EMAILS 272
　Manipulation of Evidence 274
　Climategate II 293
CHAPTER 10: THE TWILIGHT OF THE GODS 295
ABOUT THE AUTHOR 300

Introduction: The Gods Will Destroy

THOSE WHOM THE Gods would destroy, they first make mad. This ancient saying is attributed to Euripides.

Western Nations are sinking in a sea of debt, yet they are obsessed with the belief that our civilisation is being destroyed by our own actions, and the chief culprit is our need for energy. Yet, by preventing the use of fossil fuels or nuclear power, we are hastening our own departure. They are even destroying the environment they pretend to preserve as they force us to clear reserve lands to grow crops to feed us when our traditional cropland is failing is with uneconomic methods and its use for biofuel.

I have been involved with trying to understand this mad delusion for over twenty years. At first I was trapped by the authority of those publicizing the *Global Warming* theory. It was only by slow degrees that I became convinced that one aspect of its claims—one after the other—was without scientific foundation. I reached my current assessment that everything about it is a scam, a fraud, and a conspiracy, violating many principals of physics, mathematics, elementary logic and ordinary honesty.

Vincent Gray

I was never a professional meteorologist, but I ran a weather station on the roof of my school in Hammersmith, London, between the years of 1937 and 1939 and accepted, at the time, that the science of Meteorology, which represented the collected wisdom of over 200 years, was the only reliable scientific study of the climate. This scientific study has improved out of all proportion since then, and now we get the latest knowledge of the entire world climate in our weather reports every day. An alternative theory based on a postulate that changes in the climate are exclusively caused by human emissions of *greenhouse gases* has penetrated the entire academic science community worldwide. Why has genuine climate science been abandoned by so many people?

Some of the supporters of this false alternative found scientific employment from it; some were attracted by money, travel experiences and publicity, even a Nobel Prize. These reasons are easy to understand. What is difficult to understand is how so many can believe that their activity is beneficial to the *Planet* and even to human progress? It is a belief similar to a religion, yet many religions, at least in their early stages, play a part in improving human welfare. This antihuman belief hastens us on the road to disaster.

When I first developed an opposition to this aberrant pseudo scientific religion, I was almost a lone voice, both within my community and in the world outside. Now there are many groups of people who realise the absurdity of the *Global Warming* theory. Indeed, since the approved technique of assessing such warming shows that it has not been happening for the past 18 years, they promote *Climate Change*, instead.

The public is waking up to the harmful economic consequences. We may be termed *Sceptics*, but the *Warmers* call us *Deniers*, an example of the deliberately confusing use of words, with which they specialize. Nobody denies the climate is changing. Indeed it is the *Warmers* who insist that without the evil influence of humans, the climate is static and unchanging. Once evil human influences are removed by a form of World Government, they claim the climate will go back to a version of paradise.

In 2002 I published *The Greenhouse Delusion: a Critique of Climate Change 2001*,[1] a thorough analysis of the 2001 IPCC Report. There have now been two more IPCC Reports in 2007 and 2013. It is now evident that the promoters of the greenhouse theory are not just deluded—in order to promote their delusion that the Planet is being destroyed by humans, they resorted to many forms of deception, dishonesty, distortion and downright fraud to impose policies for which there is no scientific evidence. This book details how this has been done.

My autobiography, *Confessions of a Climate Sceptic*[2] describes my origins, upbringing, education and scientific career, but because I am continually subject to disbelief, insult and refusal to be taken seriously, I feel I should summarise them here. At this writing, I am 93 years old and expect this book to be my final word. I have a First Class Honours degree in Physics, Chemistry, Crystallography and Mathematics and a PhD in Physical Chemistry from Cambridge University, UK. I've had a long scientific research career in the UK, France, Canada, New Zealand and China with well over 100 peer-reviewed scientific papers on a variety of subjects, including climate. I am a Fellow of the New Zealand Institute of Chemistry. I have been an expert reviewer on all of the Reports of the Intergovernmental Panel on Climate Change (IPCC) and I am the author of 128 Greenhouse

---

[1] Gray, Vincent R., *The Greenhouse Delusion: A Critique of 'Climate Change 2001'*, Multi-Science Publishers, UK
http://www.multi-science.co.uk/greendelu.htm

[2] Gray, Vincent R., *Confessions of a Climate Sceptic*, Blurb,
http://www.blurb.com/b/1654887-confessions-of-a-climate-sceptic

Bulletins and 340 NZ Climate Truth Newsletters, most of which have appeared on the Internet.

There are now over 1350 peer-reviewed scientific papers which dispute this false theory. 31,487 American scientists signed a petition opposing it, of whom 9,029 have PhDs.[3]

This book is a detailed scientific analysis of this widely promoted climate pseudoscience. It shows how every part of this theory and its claims is the result of fraudulent manipulation of science and dishonest use of multiple forms of propaganda.

I am a firm supporter of Climate Science as embodied in the current activities of meteorologists concerned with weather forecasting, and I accept that *greenhouse gases*—predominantly water vapour—warm the climate.

But I am a *sceptic* of everything said or done by climate pseudoscience. I am a *denier* of their right to exercise control over my life and activities and those of all humans. Their unproven conclusions give them no right to influence public policy or law.

I have never been tempted to start a website as I relish the freedom to think and publish when I wish instead of being bound with the continuous task running a website involves. I was once invited to join the blog of one of my friends and I resisted it. I have learned a lot from other people's websites and blogs and have frequently contributed either articles or comments.

Matthews[4] recently made a limited (154) survey of climate sceptics who support the *The Air Vent* website and has shown that they tend to be highly educated and can be divided into several categories. Only very few are prepared to regard the global warming theory as a fraud or a scam.

---

[3] The Petition Project, http://www.petitionproject.org/
[4] Matthews, P, 2015, *Why Are People Skeptical about Climate Change?*, https://ipccreport.files.wordpress.com/2015/01/sceppre.pdf

The Global Warming Scam

The U.S. Senate[5] recently voted 98 to 1 that Climate Change is real and not a hoax.

The term *Climate Change* has acquired a dishonest double meaning: that it is caused by human emissions. Perhaps some Senators voted only for the first obvious meaning.

The word *hoax* also has double meanings. They include:

- a humorous or malicious deception[6]
- a deception, especially a practical joke[7]
- a deception, intended to trick or mislead[8]

Now, I do not think that Climate Change is humorous—it is rather more serious than a mere practical joke. Phil Jones described the *Hide the Decline* method (see Chapter 9) as a *trick*, but surely the climate change scam is much more than just a trick, it is a fraud and a *scam*, even a *superscam*.

In 1841, Charles Mackay listed historic popular delusions and crowd manias including The Mississippi Scheme, The South Sea Bubble, Tulipomania, Alchemy and the Witch Mania.[9] Climate Change should join this list. Alchemy also depended on distorted science, the belief that base metals can be turned in to gold. Another delusion was spiritualism based on the belief that it is possible to contact spirits of the dead. Both of these were

---

[5] Science, p. 201 http://news.sciencemag.org/climate/2015/01/98-1-u-s-senate-passes-amendment-saying climate change-real-not-hoax
[6] Oxford English Dictionary,
 http://www.oxforddictionaries.com/definition/english/hoax
[7] Collins,
http://www.collinsdictionary.com/dictionary/english/hoax
[8] Your Dictionary, http://www.yourdictionary.com/hoax
[9] MacKay, Charles, 1841, *Extraordinary Popular Delusions and Madness of Crowds*, Coles Publishing Company, Limited, Canada

9

taken up by scientists of the day.

Many scientists still believe in an assortment of religions quite apart from those committed to, or paid to support the environmentalist religion. Dr Rajendra Kumar Pachauri, former Chair of the IPCC, believes in reincarnation. Our entertainment industry is riddled with irrationality and primitive fantasy. Scientists forecast disasters and provide constant scares about our health. We have zombies, aliens, vampires, cyborgs, mutants, robots, ghosts, poltergeists and hobbits. Instead of news, we get a diet of murderers, accidents and the false conflict of organised sport.

Can science recover? Regardless of the nonsense, climate science prospers in the weather forecasting services, although incursions of climate change delusionists almost destroyed the UK Met Office when so many warm winters turned out so cold. The service has to cope with the influence of climate change zealots on long-range forecasting.

Large areas of science are controlled by Government bureaucrats who finance scientific justification for almost any government policy. President Nixon believed the problem of cancer could be solved merely by offering generous grants. Government funding produced a great deal of pseudoscience and plagiarism, but progress is happening.

In addition, medical science proceeds, as can be proved by the increased age of most populations. Technical development of genetics resisted attempts to kill it and the use of modern methods of producing oil has saved the fortunes of the U.S. while it can still claim to support environmental delusion.

This statement from AR5 will make them or break them:

> *Continued emissions of greenhouse gases will cause further warming and changes in all components of the climate system. Limiting climate change will require substantial and*

sustained reductions of greenhouse gas emissions.[10]

There have already been 18 years when the satellite temperature readings have been unchanged despite continued increases in emissions. How much longer will it take to show that man made climate change is simply not true?

---

[10] IPCC, 2013: *Climate Change 2013: The Physical Science Basis*, Contribution of Working Group I to the Fifth Assessment Report of the Intergovernmental Panel on Climate Change, Stocker, T.F., Qin, D., Plattner, G-K, Tignor, M., Allen, S.K., Boschung, J., Summary for Policymakers p. 17

CHAPTER 1: THE CLIMATE

DEFINITIONS OF CLIMATE include the following:

- The meteorological conditions, including temperature, precipitation, and wind, that characteristically prevail in a particular region.[11]
- The composite or generally prevailing weather conditions of a region, as temperature, air pressure, humidity, precipitation, sunshine, cloudiness, and winds, throughout the year, averaged over a series of years.[12]
- The weather conditions prevailing in an area in general or over a long period.[13]

So, *weather* means *Climate conditions which prevail in an area*, or quoting the Oxford English dictionary definition:

> The state of the atmosphere at a particular place and time as regards heat, cloudiness, dryness,

---

[11] The Free Dictionary, http://www.thefreedictionary.com/
[12] Dictionary.com, http://dictionary.reference.com/
[13] Oxford English Dictionary, http://www.oed.com/

sunshine, wind, rain, etc.

The common feature of all these definitions is that both *Climate* and *Weather* are essentially *local* and largely confined to a particular region or area.

A region or area may be quite small (there is even such a thing as a microclimate) and local topography has an important influence on wind and storm patterns.

Some climate properties are regional or even global. Changes in the sun and in the earth's axis have a global influence. Volcanic eruptions, ocean oscillations and cyclones influence several regions Local climates can be very different. It is not possible to combine all of them to provide a meaningful global average. There is no such thing as a *global* climate.

It is very difficult, even in a more or less uniform region, to derive a scientifically or mathematically acceptable average for any of the climate properties, Climate observations hardly ever comply with requirements for uniformity and symmetry which are needed to make use of standard statistical models The observations are often irregular, bimodal or skewed. Attempts to apply the popular normal distribution often claim a false uniformity or run into problems with rare events for which there are too few samples or assess regularity The term *average*, therefore, often applies only to a "*range*" or to an opinion as to what is *typical*, *normal*, or *unusual*.

Weather Forecasting

Measurement of climate properties has developed from very early beginnings with the aim of attempting to forecast its future behaviour.

Measurements are now made of a large number of properties, both near the surface and at various levels in the atmosphere using radiosondes and satellites These properties are incorporated into numerical models made up of complex equations incorporating thermodynamics and fluid dynamics of the atmosphere and the ocean.

No part of the climate is ever in equilibrium Climate is an interrelationship with a large number of meteorological and geological influences which provide a *pattern,* and for individual properties a *field* which is continually varying.

The climate depends on the behaviour of fluids in the atmosphere and the ocean. The physics of fluids involves the use of nonlinear equations with second order differential quantities demanding great accuracy in defining boundary conditions.

Edward Lorenz showed that if these methods are applied to the climate a very slight error in the boundary conditions (for example, the movement of a butterfly's wing) would be escalated by the equations, making a long term forecast impossible.[14] Lorenz concluded that for the climate *the prediction of the sufficiently distant future is impossible by any method.* There are also errors arising from the accuracy of the measurements and of the physics and the intervals in space and time of the observations.

Despite these limitations the weather forecasting services are possibly the most widely used of any scientific service. It is a necessary guide to the activity of many enterprises and individuals.

The accuracy of local weather forecasts has been recently studied by Ripley & Archibold.[15] They give a very useful summary of previous studies. They say:

---

[14] Lorenz, E., http://www.astr.ucl.ac.be/users/hgs/Lorenz-E_GarpPubl-10-06.pdf

[15] Ripley, E. A. and Archibold, O. W. http://onlinelibrary.wiley.com/doi/10.1256/wea.245.01/pdf

> *The present limit of deterministic weather predictability is a few weeks at most (Hoskins and Sardeshmukh 1987; Ripley 1988). The major limiting factors are incomplete knowledge of the atmosphere's initial state (Gilchrist1986) and imperfect understanding of atmospheric processes (Somerville 1987). The first factor is most important at short lead times while modelling errors become the dominant limitation in longer forecasts (Anthes and Baumhefner 1984).*

Their own study covers the entire Canadian system for the year 2000. They say:

> *In this analysis, we assess the accuracy of short- and medium-range forecasts of minimum and maximum temperatures and precipitation for lead times of 1 to 5 days, issued daily by the Meteorological Service of Canada (MSC, formerly the Atmospheric Environment Service) for selected cities during 2000.*

Their detailed results are tabulated. They found that temperature forecasts had a bias of about ±1°C and were rarely better than ±2°C

The results are plotted in Figure1.1.[16]

---

[16] UK Meteorological Office,
http://www.metoffice.gov.uk/aboutus/who/accuracy/forecasts

Vincent Gray

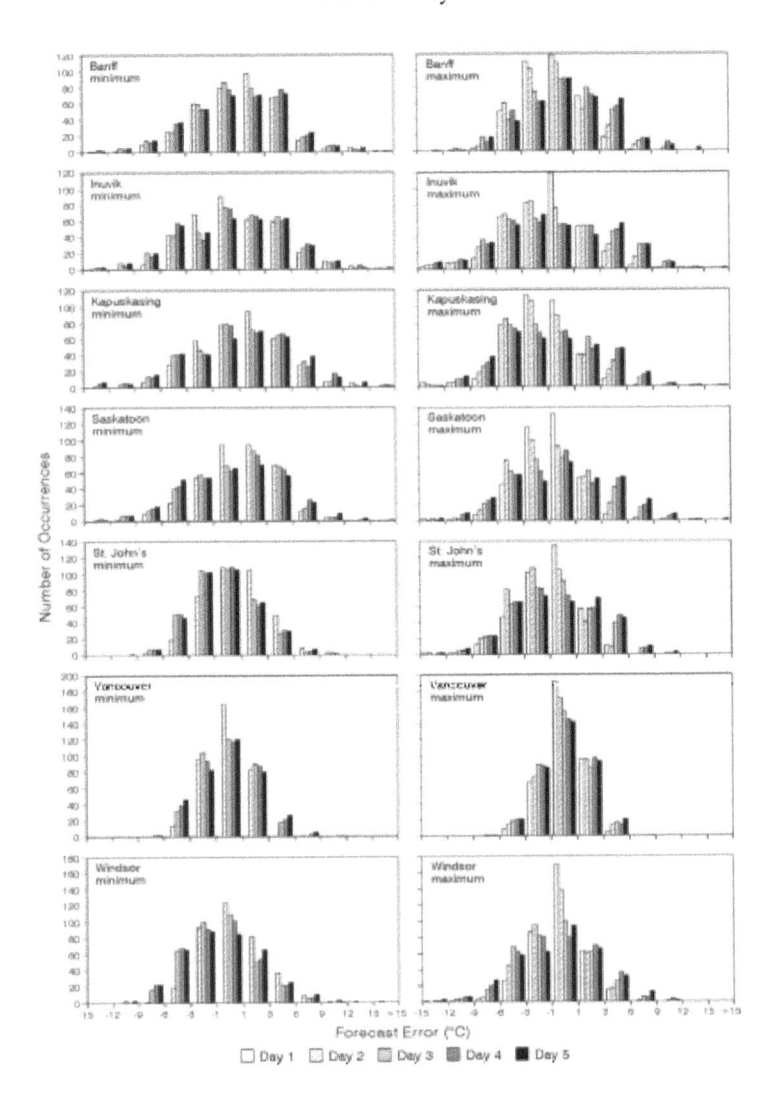

**Figure 1.1
Forecasting Errors in Different Places**

The British Met Office comes to a similar conclusion. They say:

The Global Warming Scam

Temperature
Forecasts for both maximum and minimum temperature are compared to the actual values observed at an agreed list of 119 sites across the UK. The sites used for verification are those where we have quality-controlled data and where we produce forecasts for. The early morning forecast on our website is used to produce a percentage number of the times when the forecast is accurate to within ±2°C. This is based over a rolling 36-month period to smooth out extremes and give a representative average.

Temperature forecast—performance
This information will be updated every month.
   Maximum temperature—first day of forecast
   93.8% of maximum temperature forecasts are accurate to within ±2°C on the current day (36-month average).
   Target for 2013/14 is 85.0%.
   Minimum temperature—first night of forecast
   84.3 % of minimum temperature forecasts are accurate to within ±2°C on the first night of the forecast period (36-month average).
   Target for 2013/14 is 80%.

These results are examples of current practice in advanced Western countries. Accuracy is bound to be lower in less developed countries and in the western countries in 1950, 1900 and 1850.

The accuracy of the average of a daily maximum and a

minimum temperature, the so-called *daily average*, is unknown, as it has no mathematical derivation. Any manipulations of this quantity thus has unknown and very large inaccuracies.

Climate is essentially local. Numerical climate models have to be supplemented with specific local characteristics to provide a reliable local forecast. Attempts may be made to obtain averages of individual observations such as temperature and precipitation, but it is much more difficult to average wind or air pressure patterns as influenced by more general features.

The Global Climate as an array of the climate properties at each individual and regional locality is portrayed to us every day by the press and TV weather forecasts.[17]

Since Global climate properties have to be obtained at present from an assembly of individual local and regional determinations there is no possibility of a plausible global climate model which could be capable of taking all these into account Any scientific organisation or individual scientist who claims the ability to predict the temperature change of the local or global climate beyond a few weeks, even to decimals of a degree, has to be dismissed as deluded or fraudulent.[18]

Climate as a Heat Engine

The Climate is a heat engine. Energy input is mainly short wave radiation from the sun. Energy output is mainly long wave radiation from every surface on the earth and from every level in the atmosphere, including clouds and aerosols.

Figure1.2 shows only transfer of heat by radiation. Much of the absorbed heat is distributed by conduction, convection and

---

[17] Sky News, http://www.youtube.com/watch?v=irXZqRwUQeY
[18] Global Forecast System,
http://www.ncdc.noaa.gov/data-access/model-data/model-datasets/global-forcast-system-gfs

latent heat transfer.

The whole function of the climate is to provide the energy necessary for the preservation and development of living organisms. In the process there is a complex interaction between all the climate properties.

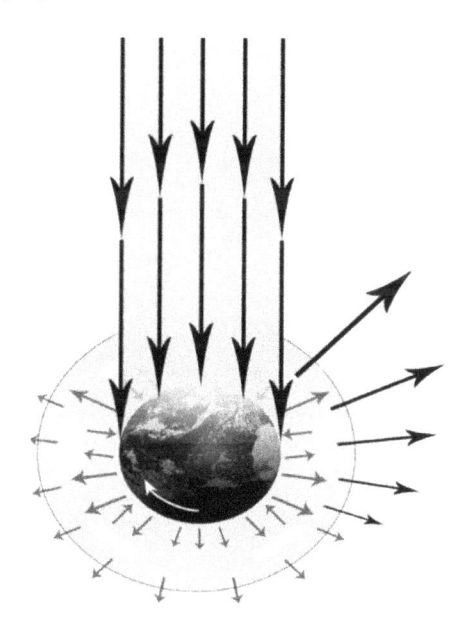

Figure 1.2
The Climate as a Heat Engine

The following diagram, which appeared in Climate Change 1995, shows the main features of the climate.[19]

---

[19] Atmospheric Circulation,
http://www.ux1.eiu.edu/~cfjps/1400/circulation.html

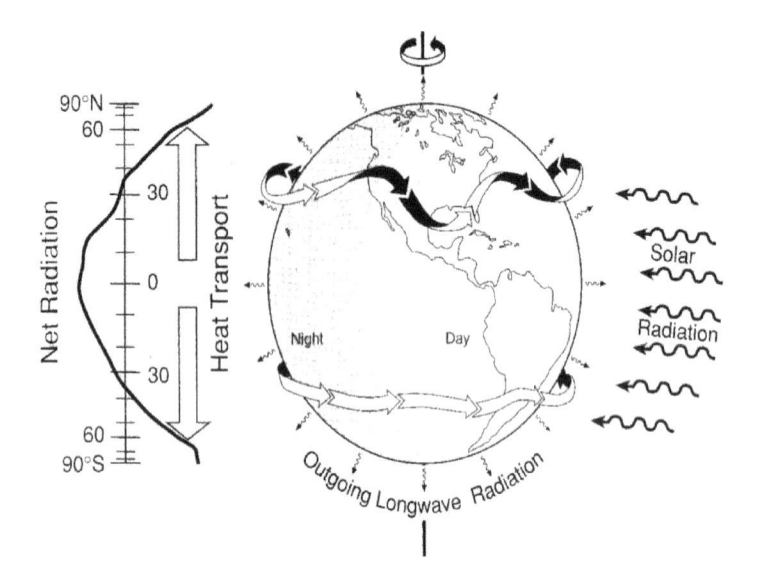

**Figure 1.3
Elements of the Climate**

Climate Properties

Observations from many instruments which are processed by the numerical climate models are obtained from a large variety of the following climate properties.

Solar Irradiance

Radiation from the sun is only received during each day, as the earth rotates on its axis and follows its orbit around the sun. It ceases altogether at night.[20] The sun passes through the

---

[20] *The Great Ocean Conveyer Belt*,
http://climatereview.net/ChewTheFat/?attachment_id=72

atmosphere at the edge of the earth and its interaction with cloud causes the effects of sunrise and sunsets.

The energy that arrives on a square metre of the earth's surface with the sun at zenith and at its average distance is referred to as the *solar constant* which has an average of 1370 W/m^2.

The actual direct solar irradiance at the top of the atmosphere fluctuates by about 6.9% during a year (from 1.412 kW/m² in early January, to 1.321 kW/m² in early July) due to the earth's varying distance from the Sun, and typically by much less than 0.1% from day to day.

Thus, for the whole *Earth* (which has a cross section of 127,400,000km²), the power is 1.740×10^{17} W ±3.5%. This amount falls to zero as it approaches each horizon and each amount actually received depends on absorption by the atmosphere, clouds and overcast.

Direct measurement of solar irradiance began with Pouillet in 1838 (see Chapter 2) and now is measured continuously by satellites. Its variability since 1610 is shown in Figure 1.4.

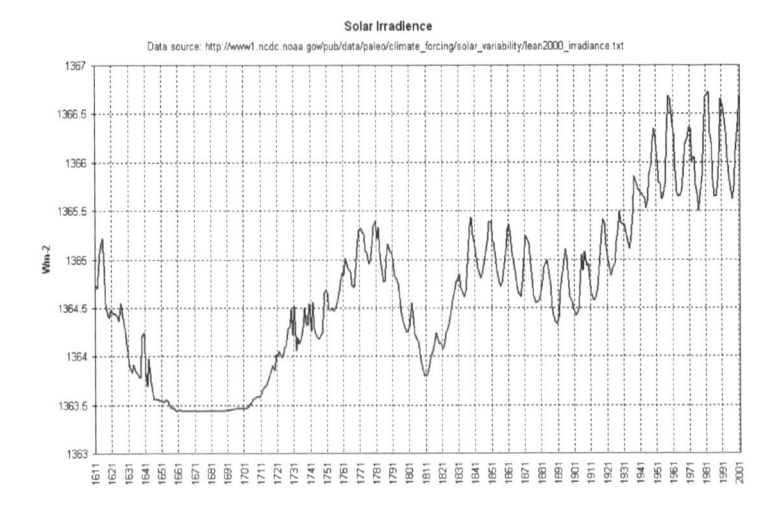

Figure 1.4
Solar Irradiance Since 1610

This graph shows the Maunder Minimum circa 1645-1750 and the Dalton minimum in 1811.

Sunspots are dark spots on the sun compared to surrounding regions. They represent concentrations of magnetic field of reduced surface temperature. Sunspots usually appear as pairs, with each spot having the opposite magnetic polarity of the other.[21] It is rather a crude way to assess solar variability but the recently revised sunspot record shows similar variability to the irradiance measurements.[22] [23]

---

[21] El Niño SOI, http://www.niwa.co.nz/our-science/climate/information-and-resources/clivar/elnino
[22] PDO Index, http://appinsys.com/GlobalWarming/PDO.htm
[23] Sunspots, http://en.wikipedia.org/wiki/Sunspot

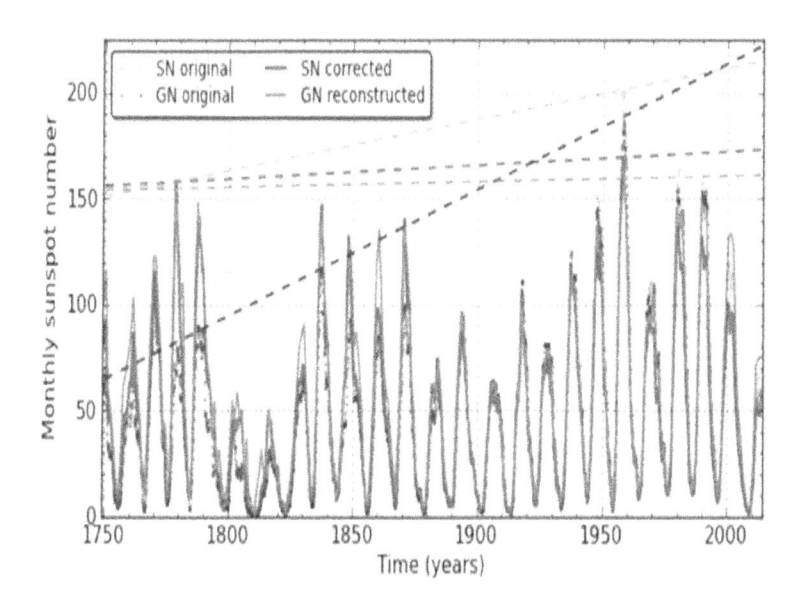

**Figure 1.5
Revised Sunspot Numbers**

The Sun contains 74.9% hydrogen which is being converted into helium by nuclear fusion heating the surface to around 6000°K.[24] The Sun then radiates energy outwards. It has the following spectral composition.[25]

---

[24] Houghton, J. T., Meira Filho, L.G., Callander, B.A., Harris, N., Kattenberg, A. and Maskell, K., editors, 1996, *Climate Change 1995: The Science of Climate Change*, Cambridge University Press
[25] *The Solar Spectrum*, http://en.wikipedia.org/wiki/Sunlight

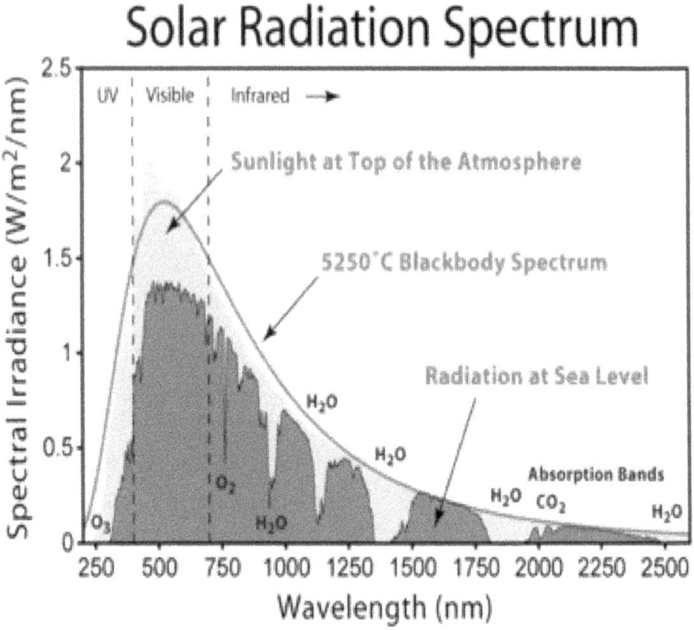

Figure 1.6
The Solar Spectrum

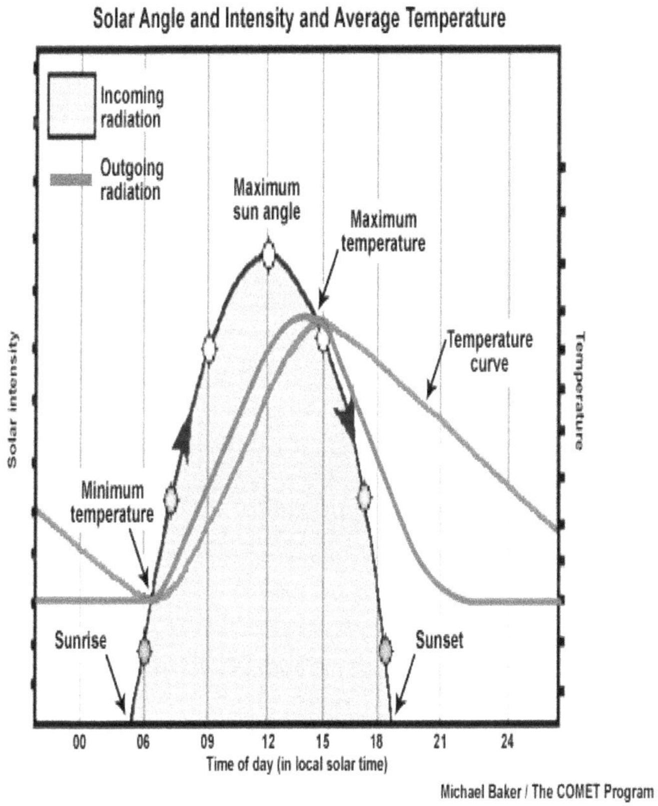
Solar Angle and Intensity and Average Temperature
Incoming radiation
Outgoing radiation
Maximum sun angle
Maximum temperature
Temperature curve
Minimum temperature
Sunrise
Sunset
Solar intensity
Temperature
00
06
09
12
15
18
21
24
Time of day (in local solar time)
Michael Baker / The COMET Program

Figure 1.7
Daily Input and Output of Radiation

The amount of radiation from the sun begins at dawn from zero to a maximum at noon and declines to zero at sunset and depends on latitude and the seasons.[26]

The radiation received follows several different paths.

---

[26] *Diurnal Temperature Variability in the Tropics* http://www.goes-thtm

Some is absorbed by the surface and is transmitted into it, depending on its local thermal conductivity. Over ice or snow some energy may form liquid water. Over the oceans, or lakes, transmission is disturbed by fluid motion in the area, both above and below the surface.

Some of the energy absorbed will transfer heat by conduction to the neighbouring layers of the atmosphere, which will rise by convection, influenced by local topology and by local wind speed and direction.

Some heat is removed by evaporation of water from damp ground or from oceans and lakes. Wind speed and direction enhances the process, so that when combined with convection the air may sometimes be warmer than the ground.

Some energy is radiated perpendicularly from the surface. It will be greatest after it has been heated by the sun and will fall progressively through the night

At night, radiation depends on heat absorbed the previous day.

Energy is radiated from every level of the atmosphere both up and down, but as it cools and the atmosphere becomes less dense the amount fall rapidly. Most radiation loss from the atmosphere takes place close to the surface, the part that causes the weather.

Atmospheric circulation distributes heat and moisture across and around the surface, reducing the difference between the tropics and the poles and up into the atmosphere as far as the tropopause.

At night, circulated air may reduce heat loss by conduction or by deposition of dew or frost.

Heat is also distributed by ocean currents which form in recognised fluctuating patterns.

Water vapour from the earth condenses to clouds as soon as it reaches the dew point. The latent heat is deposited in the atmosphere at this region, thus increasing the slope of the lapse rate.

Some clouds form rain, hail or snow which is deposited on the earth's surface, but not necessarily in the place where it came from. Since it comes from a cool part of the atmosphere it usually cools the surface on which it falls. With snow this may persist over time, abstracting further latent heat when a period of thaw ensues.

The Lapse Rate

The density of the atmosphere falls with height from reduced gravity and it cools adiabatically. The fall of temperature with height is called the Lapse Rate.

The actual lapse rate varies considerably, and depends somewhat upon the moisture content of the air, the actual change in temperature with height and the conditions of air at the surface. If the air at the surface is warmer than the lapse rate, then a parcel of air will begin to rise in the air that surrounds it. This rising air will cool off primarily by expansion at a rate known as the dry adiabatic lapse rate. When this parcel of air reaches its dew point the rate of cooling will decrease due to the heat released when water vapour is condenses to a liquid. This is known as the saturated adiabatic lapse rate.[27] The height where the parcel of air becomes saturated also is the height at which cloud starts to form. When the parcel of air can rise no higher or has released all of its moisture, the top of the cloud is reached.

---

[27] *The Lapse Rate*,
http://www.paul.moggach.yorksoaring.com/GPGSDEC11/lapse_rate.html

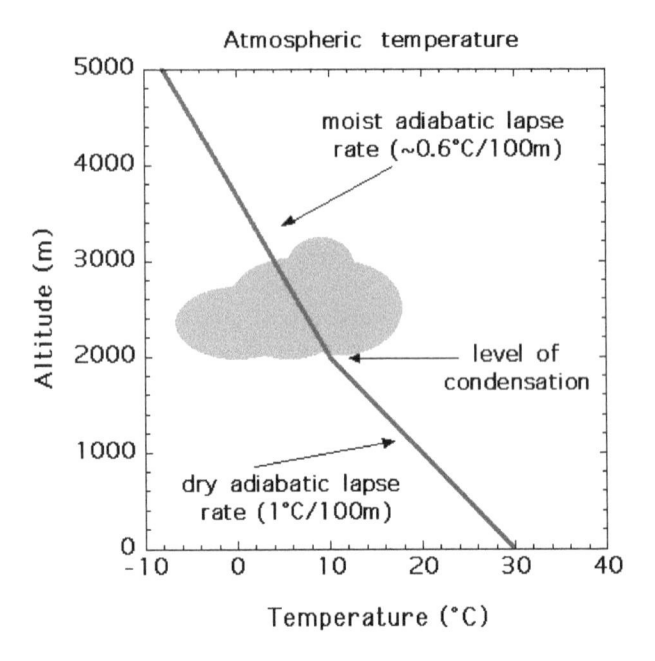

Figure 1.8
The Nominal Lapse Rate

The return of latent heat by clouds causes a decrease in this rate. In addition, there are effects from convection, radiation loss and latitude. As with all other climate properties the lapse rate profile changes continuously. It is affected by the diurnal and seasonal variability and by the general movement of the atmosphere, clouds precipitation and with cyclones and anticyclones.

Figure 1.9 shows examples of measurements made in different places.[28]

---

[28] *The Lapse Rate in Different Places*, Private Communication

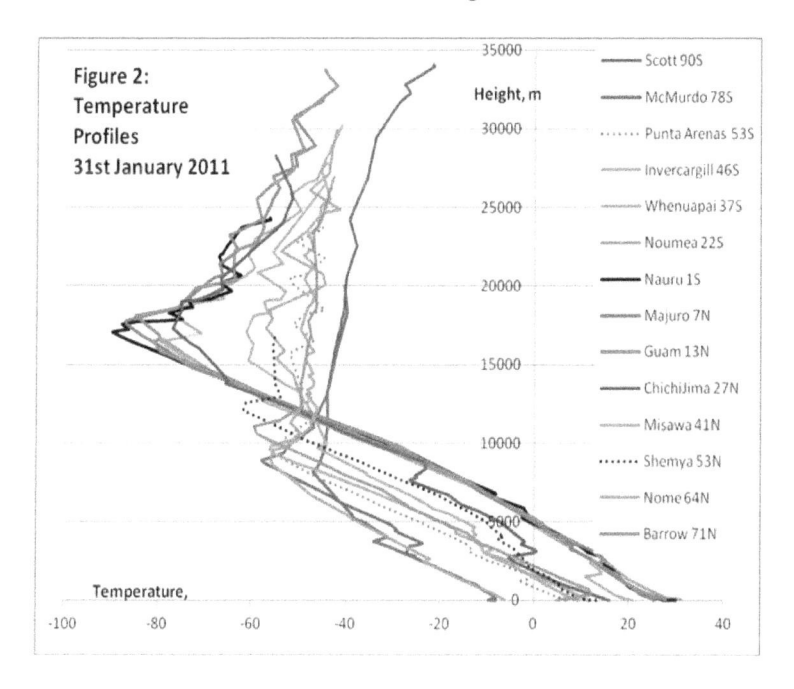

**Figure 2:
Temperature
Profiles
31st January 2011**

Series: Scott 90S, McMurdo 78S, Punta Arenas 53S, Invercargill 46S, Whenuapai 37S, Noumea 22S, Nauru 1S, Majuro 7N, Guam 13N, ChichiJima 27N, Misawa 41N, Shemya 53N, Nome 64N, Barrow 71N

Figure 1.9
Lapse Rate in Different Places

Various simplifying attempts have used the concept of a Standard Atmosphere.[29]

The surface of the earth and every level of the atmosphere radiates energy according to the Stefan-Boltzmann Law, dependent on emissivity multiplied by the fourth power of the absolute temperature. The lapse rate means that the radiated energy from the atmosphere falls very rapidly with height, so that most of it comes from close to the surface.

Radiation from the surface is upwards but from the

---

[29] *The Standard Atmosphere*, http://en.wikipedia.org/wiki/Standard_atmosphere

atmosphere it is in all directions, so that half of this radiated energy returns to the surface. By land the amount returned depends on local albedo but there is evidence that this is very low over the ocean where most is reflected.

Energy is consumed by changes in the earth's surface, by erosion, glaciers, waterfalls and the effects of cyclones and tornadoes.

Energy is used to maintain and increase all the living organisms on the earth, some of whose products can be stored for short or long periods. Stored energy from the past, such as from fossil; fuels may be restored later. Some energy may also come from within the earth with earthquakes and volcanoes or from nuclear power.

Humans modify their personal climate by erecting buildings which exclude the wind, rain, adjust temperature and lighting, and provide living and sleeping business or recreational facilities, individual design depend on local climate, availability of building materials and level of prosperity.

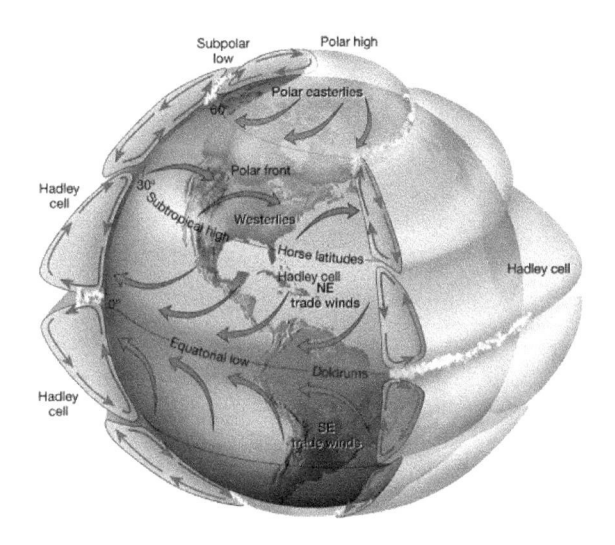

Figure 1.10
The Air Circulation System

The Circulation System

The complex patterns of the circulation of the atmosphere are the main determinants of the behaviour of the climate in every locality. The above diagram shows some of the main patterns that have been identified and classified.[30]

Air pressure determines the pattern of air circulation. The basic instrument for air pressure is a barometer. A cyclone is a rotating area of low pressure, where the flow is inward toward the centre. An anti-cyclone is the opposite, where flow is outward from the centre.

---

[30] *Atmospheric Circulation*,
http://www.ux1.eiu.edu/~cfjps/1400/circulation.html

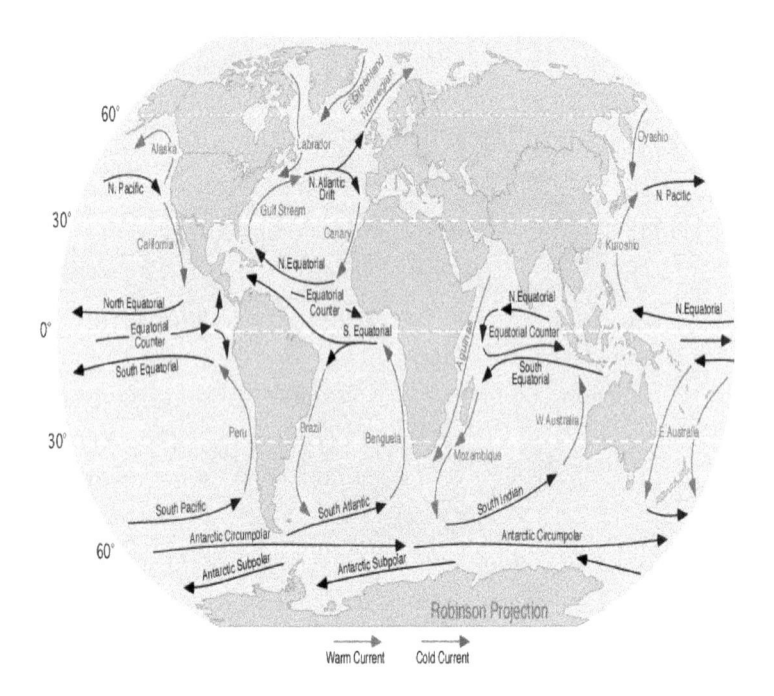

Figure 1.11
The Main Elements of Ocean Circulation

For atmospheric cyclones and anticyclones, over the northern hemisphere, air flows counter clockwise around cyclones, and clockwise around anticyclones. In the southern hemisphere, it is just the opposite.[31]

Circulation of warm and cool currents is influenced by the position of land masses and of the properties of the ocean floor. An oversimplified version of this process is The Great Ocean

---

[31] *Ocean Circulation*,
http://en.wikipedia.org/wiki/Subantarctic

Conveyer Belt as shown in Figure 1.12.[32]

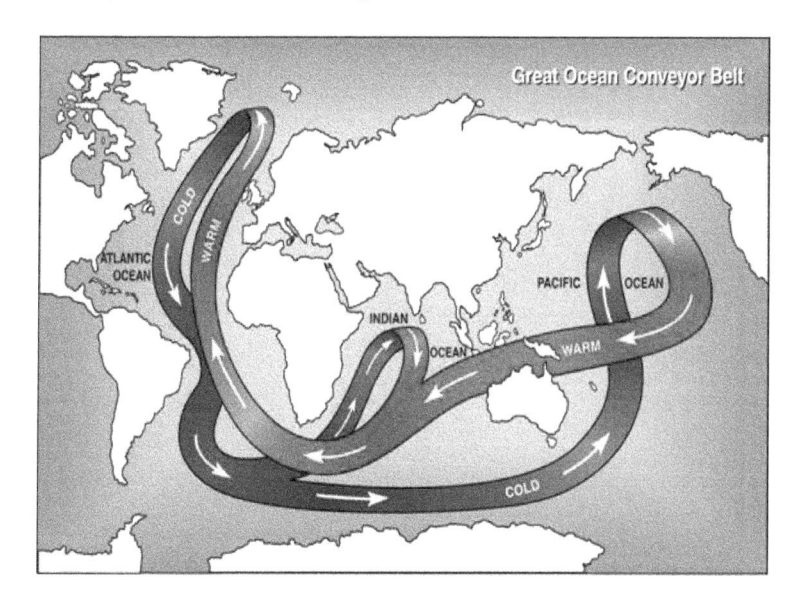

**Figure 1.12
The Great Ocean Conveyor Belt**

Ocean Oscillations

The principal oscillations are:

- El Niño/Southern Oscillation (ENSO) which is observed in the southern Pacific and has a periodicity of 3 to 8 years.[33]

---

[32] *The Great Ocean Conveyer Belt*, http://climatereview.net/ChewTheFat/?attachment_id=72
[33] PDO Index, http://appinsys.com/GlobalWarming/PDO.htm

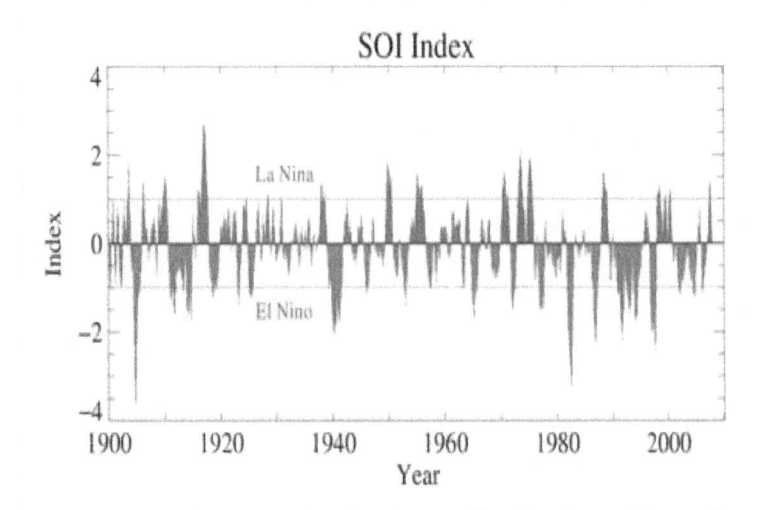

Figure 1.13
The El Niño Southern Oscillation Index (ENSO)

- Pacific Decadal Oscillation (PDO) which is observed over the whole Pacific hand has a periodicity of two or three decades.[34]

---

[34] *El Niño SOI*,
http://www.niwa.co.nz/our-science/climate/information-and-resources/clivar/elnino

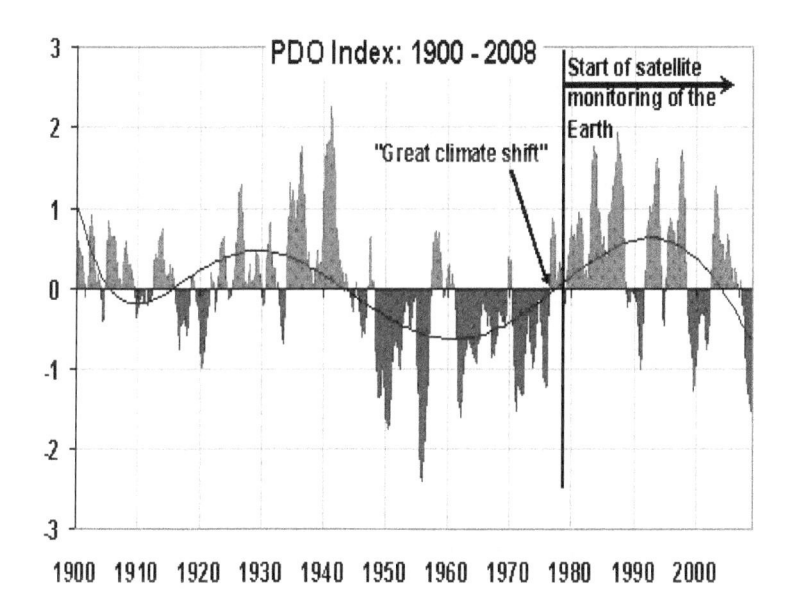

**Figure 1.14
Pacific Decadal Oscillation Index**

- *North Atlantic Oscillation (NAO)* which is observed over the northern Atlantic Ocean and has a period of around one decade

Other oscillations affect the Arctic, Antarctic and Indian Oceans.

ENSO has a global effect on the climate. The *PDO* influences the Arctic ice cover.

Meteorology Today

There are now some 8,000 official weather stations measuring a very large quantity of local and international climate properties and processing them to supply weather forecasts, which now apply to the entire lower atmosphere, the surface and the oceans. Sophisticated world maps with temperatures for three times a day, air pressure. Wind circulation, wind velocity, precipitation, relative humidity, hours of sunshine, behaviour of clouds, and many other features are presented on television and newspapers all over the world on a continuous basis.

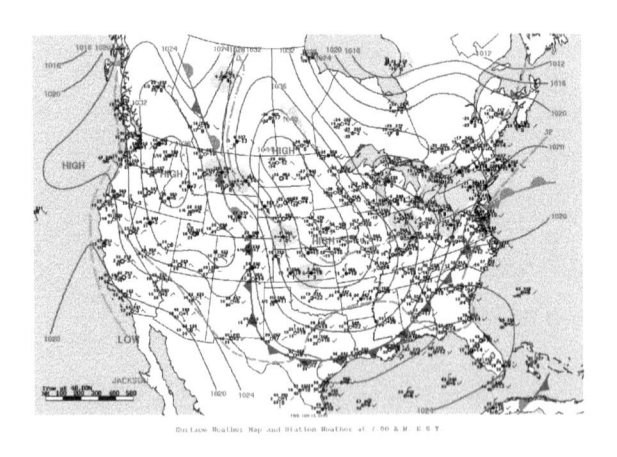

**Figure 1.15
Typical Weather Map**

This service is by far the most widely used scientific operation, for individuals, communities, businesses and nations.

Figure 1.15 shows a typical weather map of North America showing air pressure and wind direction.[35]

---

[35] National Weather Forecast Office, http://www.erh.noaa.gov/btv/events/15Jan2009/wx.shtml

Other maps show temperature measurements, often three times a day, the progress of cyclones and many other properties on different scales. Nowadays many are animated.

Climate science is regional. Each region has climate models which are influenced strongly by the significant local climate effects in that region.

There is, at present, no prospect of an overriding climate model suitable for the entire climate and any claim that such a system exists is spurious.

Climatology

Meteorology is the discipline mainly concerned with forecasting. *Climatology* is a related discipline which uses climate measurements and historical and geological information to assess information about the likely general climate behaviour in a particular place, region or historical or geological epoch.

It is to be regretted that some climatologists have attempted to extend their discipline into attempts to forecast future global climate. So far these attempts have had no useful influence on weather forecasting and no successful future global prediction has ever emerged. Future Chapters will explain why this is so.

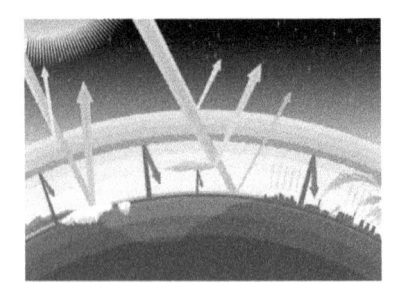

CHAPTER 2: ENVIRONMENTAL RELIGION

The Origins of Religion

THE HUMAN INSTINCT to worship one or more nebulous beings in nebulous localities goes back to the five million years of our early evolution when we were faced with hostile surroundings and the necessities of daily survival. Societies were disciplined and dictatorial, much as monkeys, deer, lions and meerkats are today; and some sub-groups of contemporary human society still exist today—such as armies or criminal gangs. Obedience to a leader and rejection of those who do not conform is essential. Those who survived were the best followers of orders.

In order to ensure obedience to the leader, rituals were developed to make him appear special. He was approved and advised by supernatural beings, and the people had to be made to believe a whole theology to support the legitimacy of the leader. Priests and a church which sanctified the leadership were supplied. Anyone who disagreed was rejected, and their genes did not survive. The survivors evolved an instinct for worship and acceptance of authority which many still possess.

In this way, over millions of years, we acquired an instinct to believe and accept irrational and supernatural explanations for the parts of our life we cannot control. We are bred to worship and support religious leadership.

The world today is still controlled to a great degree by religious beliefs, which assist the maintenance of elite leaders. Such systems tend to be resistant to innovation, to changes in technology and new forms of social organisation. History holds many examples of the downfall of entire civilizations because of religious and social dogma—conquered, replaced or overcome by those who have better technology and less rigid conformity.

The most prosperous and civilized nations of this world have become this way because they were able to overcome the stultifying influence of religions and the backwardness they encourage. Although no country is free from religious influence, in one way or another, many of us have minimised this influence, so that we can develop new technologies and new ideas which can promote economic and social progress despite religious interference.

Many of the most prosperous states in the world no longer have the burden of an official state-sponsored religion forcing citizens to behave in the way determined by the priests.

Humanist and sceptical groups in many countries established the right of freedom of worship, or of non-worship. New Zealand is one country where a large proportion of the population declare themselves to possess no religion, where Members of Parliament and those giving evidence in Court are not forced to swear allegiance to any God.

At least, this was so until the growth of the *Environmental Movement* in the second half of the 20th century. This movement has now reached the dimensions of a State and World Religion, both in New Zealand and in the world at large.

It is difficult to draw distinctions between religions, a cults and irrational beliefs. All have their origins in the human instinct to worship. They are just a matter of degree. Nazi Germany fell afoul of the belief that Germans were a superior race. In Stalinist Russia any measure could be justified on behalf of a mythical *working class*. These beliefs were every bit as strong and as disastrous for their victims and their believers as any belief based on a supernatural God.

Environmentalism has a God, a Heaven and a Devil, so it comes close to qualify as a religion. It demands constant and increasing sacrifices and represents an increasing threat to freedom of thought, of science and of future human progress. It has Ministers in every country but does not have a Pope or a single Holy Book…

Early humans were surrounded by a hostile world into which incursions were only made for essential purposes: food clothing and shelter. Industrial developments in the 18th century led to a situation where the outside world was no longer dangerous, as humans had guns, provisions and transport.

The *Romantic Movement* reversed the dangers of the outside world to argue that a world without humans was actually superior and artistically sublime. The concept of *Nature* which was superior to human habitation became widely accepted. Even the most primitive humans came to be regarded as superior (*Noble Savages*) from those like Rousseau, who hated humans and human society.

The concept of *The Environment* has its origins in the belief in *Nature*. Traditionally, humans have always considered themselves to be superior to other creatures.

Members of the Romantic Movement, who rebelled against the pollution and squalor of the Industrial Revolution, considered humans were inferior to other organisms which existed in a separate part of the world called *Nature*. Everything that happened there was *Natural* whereas everything the humans did was *artificial* and thus inferior.

The prototype science fiction novel Mary Shelley's

The Global Warming Scam

Frankenstein established a genre which still dominates our literature and public entertainment with its message that science and technology is dangerous, harmful and will inevitably cause disaster.

The Environment carries this belief much further, to the extent that *Environmentalism* has become a substitute for religion. Everything that humans do is harmful to *The Environment*.

The Environment has not only become a substitute for God. It is also a Heaven on Earth. There are no holy scriptures, but there is an amateur priesthood who are consulted regularly to comment on daily news items. In each country there is a Minister and Ministry of the Environment which promote its demands. There is a United Nations Environment Programme. There are many local and international organisations which promote its doctrine. These include Greenpeace, The Worldwide Fund for Nature, The Sierra Club, Friends of the Earth, and many others.

The demands of *The Environment* include regular and increasing sacrifices on its behalf. All human activity is either forbidden or it has to be more expensive than before. It has to endure uneconomic solutions and expensive permission from environmental authorities.

Where is *The Environment?* It is everywhere and nowhere. What is it? It is a mixture of Heaven, Nature, Never Never Land, Narnia, Erewhon and Utopia. One thing it cannot be is a territory where humans have no influence, just as there is no territory which has no influence from other organisms. We must continually protect the *Environment* from the harm done by humans or at least *Mitigate* them. It would perhaps be better if there were no humans at all.

There are several dogmatic beliefs of this substitute religion which are in conflict both with scientific thought and straightforward commonsense.

Most of us decide the merit of alternative actions by *Risk Analysis* where risks and costs of alternatives are assessed and a choice is made on the level of risk compared with estimated rewards. For public bodies this takes the form of *Cost-Benefit Analysis*. *Environmentalism* replaces this rational process with the *Precautionary Principle* where the supposed benefit to the environment is always chosen, even if the risk is zero. This is used to enforce costs, restrictions, and threats to human life.

Most religions claim that they exist to benefit humans in general, although believers usually have priority. *Environmentalism* is the only religion where humans are a *danger* to the *planet* and must not just be brought under control, but maybe even eliminated altogether, in order to benefit all other organisms, particularly those who are *endangered* by the mere existence of humans.

The most outrageous demands of most religions are tempered either by the existence of a big boss (for example, the Pope) or by a holy book which supposedly summarises their doctrine, but which is in practice subject to interpretation.

Environmentalism has no such traditional authority. Any person claiming to be an *environmentalist* or an *ecologist* can claim the right to comment, object to or forbid any action, and is currently welcomed by the media while rarely permitting a dissenting opinion.

The world's news services are trawled to reveal any event or behaviour which can be construed as evidence of damaging human behaviour. Everything we do is unsafe or harmful. It is surprising that we seem to be living longer in spite of the *threats* that surround us.

Every event is a disaster and every potentiality is based on a combination of the worse circumstances that can be envisaged.

It is difficult to characterise this attack on rationality. Traditional religions are tolerated in most Western nations. We may object to *discrimination against gays, arranged marriages, male or female circumcision*, but in general, most of us favour the right to freely practice religion and express religious opinions.

It is different if, as with the environmental religion, they have political parties and policies which interfere with freedoms of the entire population. Even worse is when they employ scientists to promote what are essentially political and religious policies. They use the public reputation of science and scientists to imply truthfulness, balance and impartiality to enforce biased political policies which are expensive, harmful and dangerous. They employ an array of devices to conceal their true objectives. This is not just delusion or misunderstanding, it is true and fraudulent.

Evolution Science

The claims of Environmentalism are in contradiction to the science of evolution.

The Greek philosopher Heraclitus is supposed to have said *everything flows* and for most people it is surely obvious that everything about the world is constantly changing.

Medieval society believed otherwise. They thought everything about the world was static and unchanging, apart from the occasional cyclones, earthquakes and volcanic eruptions, often blamed on the anger of God, When these special events were over, the world returned exactly to where it was before.

Carl Linnaeus, who in 1735 launched the most popular classification system for all organisms, believed that they were all constant and unchanging, individually created by the almighty.

It was only early in the 19th century with the discovery of fossils of extinct animals and the geology of Hutton and Lyell that it became obvious that the earth and all of its organisms have changed by evolution.

Early works of Cuvier, Erasmus Darwin (Charles Darwin's grandfather), Lamarck and Robert Chambers led to the formulation of a general theory of the mechanism of evolution by Darwin and Wallace in 1858 and the publication of Darwin's book *The Origin of Species* in 1859. This theory has been comprehensively confirmed by recent work on genetics and the mechanism of heredity.

Darwin noted that most organisms had more offspring than could survive. They also possessed variability. The survivors of each generation would be those from this variety that could cope best with each changed circumstances. Over time the living organisms would undergo observable change. Each organism is involved in a *struggle for existence* or *survival of the fittest*. Darwin showed that this process could be carried out on domestic animals and birds by selective breeding.

Darwin himself was captured by his community beliefs. His wife was an enthusiastic Christian and he struggled long before admitting in his final autobiography that he was an *Agnostic*. The confession was censored from this work until his granddaughter published the full version in 1958.

Darwin saw that the process by which the *fittest* survived was *Selection*, but he still believed in *Nature*, so he made a distinction between *Natural Selection* and *Artificial Selection*, admitting a human privilege that his theory denied.

The process does not need this distinction. *Selection* operates in the same manner whether it is by deliberate or inadvertent action of a particular organism, an action of the climate, or earth changes. They all have the effect of processing evolution.

Evolution by selection applies to all organisms including humans. Humans therefore have no privileges over others. They have no supervising deity, no after life, and no responsibility for other organisms except for their own survival.

Darwin's theory was later wrongly altered to incorporate

the idea that humans are special by Julian Huxley, Ernst Mayr and others.

The Origin of Species explained that the term *Species* is merely a classification category decided arbitrarily by taxonomists. The decision on separation of Species is both arbitrary and variable. At present there are two competing methods, that of Linnaeus, based on appearance and behaviour, and that based on DNA sequences. Whichever is chosen, the number of species can be any figure that taxonomists decide; there is no definite number.

The reasons for separating one species from another are not standardized and are almost impossible to decide for fossils viruses or bacteria.

Organisms participate in mutual struggle for existence, and some are more successful than others. The process is so complex that the future cannot be predicted. Occasionally a particular organism becomes extinct and human activity may play a part in the process, but we are not obligated to intervene unless our own interests are involved.

The disagreement between the Environmentalist Religion and Evolution Science can be summarised in the following table.

ENVIRONMENTALISM	EVOLUTION
The Environment	The Environment has no physical existence. Organisms interact in such a complex manner that no clear boundary can be drawn between one group of organisms and another.
Humans are responsible for the *planet*.	Humans are just one of many organisms interacting with one another. No organism is responsible for any other

	organism unless its own interests are involved.
Humans are destroying the *planet*.	Humans have only a minor influence on organisms in general.
Sustainability	Sustainability is the reverse of evolution. Evolution cannot be stopped, let alone reversed. Each organism tries to take advantage of evolution, but to stand still is the road to decline.
Endangered Species	All organisms evolve and new ones replace old ones. The term *species* is a human artifact which tries to classify evolutionary progress. Organisms that become less common make way to others which renew the world.
Ecosystems	Ecology is the study of interaction between organisms. It is a dynamic process and it occurs at every level. It is not possible to divide organisms into convenient packages. Every region, whether small or large, is constantly changing, never static.
Biodiversity	The numbers of organisms interacting in one area cannot be satisfactorily counted, and they depend on the climate and the evolutionary success of each organism. There is no ideal

	number nor is a larger or smaller number a reason for concern.
Conservation	Possible only to a limited degree. No point in opposing inevitable change.
The Selfish Gene	A theory that only the evolution of individuals is important. It ignores the importance of social evolution and the development of instincts.
The Noble Savage	The belief that primitive peoples are somehow superior and free from problems of more developed societies is untrue.

People are, whether they realise it or not, controlled by evolution. Those individual or collective actions which assist not only future survival but also the development of a harmonious society are unpredictable.

The past behaviour patterns of humans must be judged to have succeeded since the numbers and influence of humans have increased, despite actions such as war, genocide, famine, earthquakes, volcanoes and changes in the climate. Human use of science, technology, innovation and social interaction have brought about these changes and they must undergo continuous encouragement and development for social or individual survival in the future.

Environmentalists are opposed to all technological evolution. If they succeed in preventing it, they will cause the downfall and disappearance of the human race, something they seem so fond in anticipating.

The most basic of all instincts is selfishness without which none of us can survive. A harmonious society has to invoke a whole array of procedures to control individual or collective selfishness and eventfully to breed generations where collective benefit is instinctive.

Environmentalists use science to exaggerate every disaster and use science to try to claim they will get worse.

In order to understand how their deception is made it is necessary to understand the *scientific method*.

The Scientific Method

The Oxford English Dictionary defines the *scientific method as:*

> *a method or procedure that has characterized natural science since the 17th century, consisting in systematic observation, measurement, and experiment, and the formulation, testing, and modification of hypotheses.*[36]

The Free Dictionary:

> *The principles and empirical processes of discovery and demonstration considered characteristic of or necessary for scientific investigation, generally involving the observation of phenomena, the formulation of a hypothesis concerning the phenomena, experimentation to demonstrate the truth or falseness of the hypothesis, and a conclusion that*

---

[36] *The Scientific Method*, http://www.oxforddictionaries.com/definition/english/scientific-method

validates or modifies the hypothesis.[37]

Induction

The procedure described in these definitions is called induction. Its use as the scientific method was publicised by Francis Bacon.[38] The inductive method procedure may be shown diagrammatically as shown in Figure 2.1.[39]

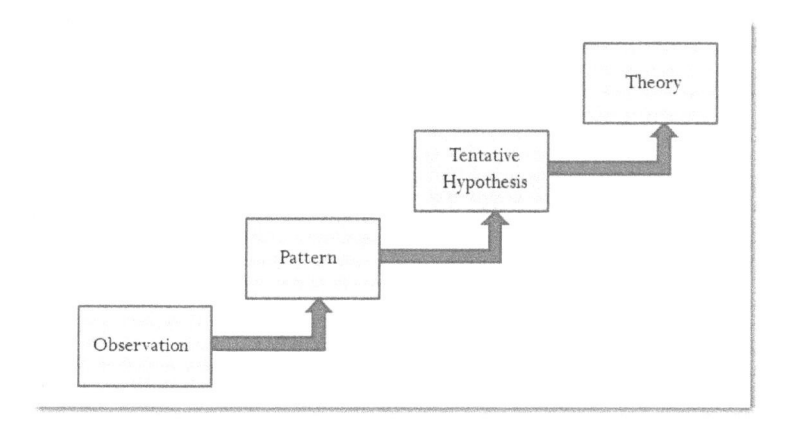

**Figure 2.1
The Inductive Method**

Observation comes first. The hypothesis and then the theory arise from the observations. The validity of the theory depends on the efforts placed in its modification from future observations.

---

[37] *Knowledge Base*, http://www.socialresearchmethods.net/kb/dedind.php
[38] *Francis Bacon,* http://www.iep.utm.edu/bacon/
[39] *Knowledge Base*, http://www.socialresearchmethods.net/kb/dedind.php

For most of us the scientific method is what is described in official scientific publications. Yet P.B. Medawar in his *Is the Scientific Paper a Fraud?* argues that:

> *The scientific paper in its orthodox form does embody a totally mistaken conception, even a travesty, of the nature of scientific thought.*
>
> *The conception underlying this style of scientific writing is that scientific discovery is an inductive process. What induction implies in its cruder form is roughly speaking this: scientific discovery, or the formulation of scientific theory, starts with the unvarnished and unembroidered evidence of the senses. It starts with simple observation—simple, unbiased, unprejudiced, naive, or innocent observation—and out of this sensory evidence, embodied in the form of simple propositions or declaration of fact, generalizations will grow up and take shape, almost as if some process of crystallization or condensation were taking place The theory underlying the inductive method cannot be sustained. Let me give three good reasons why not. In the first place, the starting point of induction, naive observation, innocent observation, is a mere philosophic fiction. There is no such thing as unprejudiced observation. Every act of observation we make is biased. What we see or otherwise sense is a function of what we have seen or sensed in the past.*[40]

---

[40] Medawar P B, 1964, *Is The Scientific Paper a Fraud?*, http://www.albany.edu/~scifraud/data/sci_fraud_2927.html

The Global Warming Scam

David Hume and particularly Karl Popper have also asserted that this procedure is invalid.

Popper says:

> By an inductive inference is here meant an inference from repeatedly observed instances to some as yet unobserved instances, I hold with Hume that there simply is no such logical entity as an inductive inference; or, that all so-called inductive inferences are logically invalid. I agree with Hume's opinion that induction is invalid and in no sense justified.[41]

Popper is quite famous for the following quotation:

> ...no matter how many instances of white swans we may have observed, this does not justify the conclusion that all swans are white.

Of course, he is right. The statement that *all swans are white* is incorrect *today*. If it follows from an inductive study of swans it is obviously incomplete. It could not be considered to be a conclusion from a properly conducted scientific study. It is easy to *falsify* it.

If it was made before the discovery of Australia, it was correct and could be acceptable as part of a properly conducted scientific study of swans at the time.

Scientific conclusions are never final, as future scientific discoveries may modify or alter them. It is therefore foolish to suggest that some past scientific conclusions are false.

---

[41] Popper, K. R., *The Problem of Induction*, http://dieoff.org/page126.htm

Science does not claim truth. Scientific statements are always temporary, because they always assume that they may need to be replaced or modified in the future.

Popper makes an excuse for his hostility to induction in the following passage:

> *What we do use is a method of trial and the examination of error; however misleadingly this method may look like induction, its logical structure, if we examine it closely, totally differs from that of induction.*

He also makes it plain that he prefers deduction:

> *I assert that scientific knowledge is essentially conjectural or hypothetical.*

Deduction

The alternative logical procedure Popper seems to favour is deductive reasoning.[42]

---

[42] Popper, K. R. 2010 (1959), *The Logic of Scientific Discovery*, Routledge

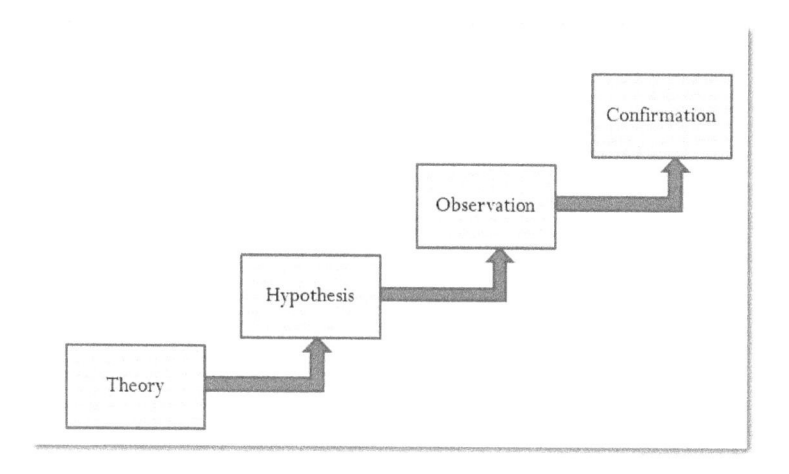

**Figure 2.2
Deductive Reasoning**

Here the study begins with a proposed theory and the investigation consists of an attempt to find observations and make experiments which might confirm the theory.

Medawar is equally scathing about this system:

> ...deduction in itself is quite powerless as a method of scientific discovery—and for this simple reason: that the process of deduction as such only uncovers, brings out into the open, makes explicit, information that is already present in the axioms or premises from which the process of deduction started. The process of deduction reveals nothing to us except what the infirmity of our own minds had so far concealed from us.[43]

---

[43] Ibid

Here he seems to be saying that we did not need to be told that the earth goes round the sun or that an atomic bomb can be made because we ought to have known it already. This is surely nonsense.

Harré, in his book on *Great Scientific Experiments*,[44] gives a useful discussion on deductive and inductive logic and shows that his examples often use a mixture of both deductive and inductive reasoning.

Harré gives the following examples of inductive reasoning:

Galileo *The Law of Descent.*
Robert Boyle *The Spring of Air.*

Many more may easily be found:

- *Darwin*'s theory of evolution arose from observations.
- *Alfred Russell Wallace* went through the same process.
- Many geological theories arise from observation. *Continental Drift* is an example.

Biology has many examples of induction:

- *Harvey*. Circulation of the blood.
- *Fleming*. Discovery of penicillin.

Then in Physics we have:

- *Newton*'s apple.
- *Becquerel* radioactivity.
- *Hahn* Nuclear Fission.

---

[44] Harré, R., 1981, *Great Scientific Experiments*, Phaidon, Oxford

The prevalence of deductive reasoning hardly needs mentioning. We have been the beneficiaries of Newton's Laws of Motion ever since he proposed them. With thermodynamics they control our knowledge of the atmosphere.

Popper and Medawar were wrong to condemn induction in science and Medawar was wrong to condemn deduction. Both of them implied that science is exclusively personal. In fact, science is a cooperative enterprise where the different necessary procedures of scientific investigation may be made by different individuals, or even by different disciplines.

Science builds on past discoveries. Newton claimed that his discoveries were obtained by *standing on the shoulders of giants*.

Validation

Any scientific study, theory or conclusion should be *validated* before it could be generally acceptable, it should be capable of simulating past observations and of forecasting all future conditions for which it may be considered appropriate, both to a satisfactory level of accuracy.

Falsifiability

Popper considered that scientific statements can only be accepted if they can be falsified:

> *There can be no ultimate statements in science: there can be no statements in science which can not be tested, and therefore none which cannot in principle be refuted, by falsifying some of the*

and...
> conclusions which can be deduced from them.[45]

> In so far as a scientific statement speaks about reality, it must be falsifiable; and in so far as it is not falsifiable, it does not speak about reality.

All scientific observations should be capable of being repeated by an independent investigator.

There are considerable difficulties in applying the test of falsifiability to investigations outside a laboratory. Such studies therefore need to be validated.

Opinions of Experts

A number of scientific disciplines are unable to apply validation, or qualify for falsifiability. They include geology, taxonomy, cosmology and anthropology. It is also true of disciplines on the edge of science, such as psychology, sociology and economics.

The opinion of experts could be criticised from the fact that the experts concerned can often be considered to have a conflict of interest since they benefit from income, prestige or status connected with their expertise.

Mathematical Models

Scientific observations may most conveniently be summarised by their ability to resemble a mathematical model. It is important to recognise the distinction between science and mathematics.

Mathematics is formalized logic. It is based on deriving the consequences of rigidly defined assumptions. It is never

---

[45] Popper, K. R., *Science as Falsification*, http://www.stephenjaygould.org/ctr

necessarily a representation of reality, but its conclusions are truthful.

Scientific theories often use mathematical models to summarize observations or measurements.

An example is the model the Greek astronomer Ptolemy made of the movements of the planets using the earth as its centre. It involved complex *epicycles* for the planets and an *equant* for the earth.

When Copernicus proposed that planets and the earth revolved around the sun, Galileo, Kepler and Newton provided a much improved mathematical model for this theory which was much simpler. Unfortunately, some people believe the model rather than reality. So, some schoolteachers and scientists think the sun does not go round the earth. Yet everybody else knows it does. The principle of relativity had to await Einstein before it was used to modify Newton's Laws and even he did not supply a relativity theory for rotating bodies.

Accuracy

All measurements are inaccurate but guidance on which of several measurements is better only came with the development of mathematical statistics models which could simulate the behaviour of a set of measurements. Although there are now many such models intended for different purposes only one of these is currently in regular use. It is the Normal or Gaussian distribution and it is present on every computer spreadsheet and *scientific* calculator. It is often successful for sets of measurement that are approximately uniform, symmetrical and numerous but it unsuitable for skewed distributions and for outliers.

Climate Science

Measurements of the climate are essentially non-reproducible and therefore incapable of being falsified. Weather Forecasting is largely a deductive process using local numerical models compiled from basic physics and local climate pattern development. It is comprehensively validated from the success of its forecasts.

Climatology

An example of a scientific discipline which cannot be falsified or validated is *climatology*.

Its results are based on the opinions of experts.
According to Merriam-Webster, Climatology is:

> *The scientific study of climates*[46]

...and according to Dictionary.com:

> *The science that deals with the phenomena of climates or climatic conditions*[47]

As Chapter 1 shows, meteorology is also *the scientific study of the climate*, much of which has to be taken into account when making a forecast. Climatology is much concerned with past climates for which the only *average* is often the *best guess* of *experts*.

---

[46] *Climatology*, http://www.merriam-webster.com/dictionary/climatology
[47] *Climatology*, http://dictionary.reference.com/browse/climatology

The Global Warming Scam

Climate Change Pseudoscience

It is only recently that climatology has been overwhelmed by intrusion of the doctrine of climate change which does purport to forecast future climate, but, so far, unsuccessfully, by defying so many accepted principles of traditional science to an extent that it has to be considered as pseudoscience. It has only been able to survive by suppression of all forms of open debate and by political pressure.

Climate Change Pseudoscience is a deductive process, designed to provide evidence for the theory that climate is controlled by emissions of trace atmospheric gases (so called greenhouse gases) by human activity.

Attempts to validate it have failed. It has been repeatedly falsified. It thus qualifies as a *pseudoscience*.

It has also rejected much of the established knowledge of climate science.

ASSUMPTIONS OF CLIMATE CHANGE PSEUDOSCIENCE	ASSUMPTIONS OF CLIMATE SCIENCE
Emissions of greenhouse gases warm the earth.	Most scientists agree that this is so. But there is no evidence of harm and good evidence that they are beneficial.
The Earth is flat.	The Earth is an oblate spheroid.
The Sun shines both day and night with equal intensity.	The Sun shines only by day, varying with time and season and affected by clouds overcast and precipitation.
Radiant energy entering is balanced by energy leaving.	No part of the earth is ever in equilibrium and there is never an energy balance.

No energy is used to maintain life, yet humans emit gases.	Entry energy could never equal that leaving because energy is used to maintain life and erode the earth. Storage may change.
All heat transfer is by radiation.	Most atmospheric heat transfer is by conduction, convection, and latent heat transfer.
Temperature of the earth can be assessed by multiple manipulation of weather station and sea surface measurements.	It is impossible to measure the average temperature of the earth by any method.
Temperature is increasing decadally.	Annual temperature change by the IPCC method has been unchanged for 18 years.
Models are evaluated by scientists with a conflict of interest.	Distortion, manipulation, fabrication and fraud have been tolerated.
The process is claimed to follow the scientific method.	No claim has ever been validated and they have been repeatedly falsified.
Models never make predictions, only projections.	Weather forecasting models predict temperature for a few days ahead to ±2°C with a bias of ±1°C.
Greenhouse gases are well mixed.	Greenhouse gases are never well mixed as shown by NASA satellite measurements.
The sea is being acidified.	Parts of the ocean are already saturated as they emit carbon dioxide. Extra carbon dioxide

The Global Warming Scam

	would increase their proportion slightly.
Sea level is rising.	Most measurements are upwardly biased from storm damage, harbour improvement, land subsidence from buildings and removal of minerals and ground water. Since GPS levelling was established, changes have been small or negligible.
Ice is melting.	The Arctic ocean periodically changes with currents The ice on the Antarctic continent is currently increasing.
Extreme events are increasing.	Evidence is poor.

CHAPTER 3: ENVIRONMENTAL SCAMS

IN ORDER TO impose dogmas and restrictions on the world, the various environmental movements abandoned rational discussion, scientific method and open debate and adopted the principle that the ends justify the means.

Distortion and fabrication of evidence has become routine, and is a feature of our news bulletins, scientific journals, schools and university departments. Sceptics are called *deniers*, hounded from employment and generally publishable only on the Internet.

Exaggeration, partial evidence, speculation and unjustified assumptions are compounded, often with the help of computers, to provide scare scenarios for the future. A useful summary is available at Environmental False Alarms.[48]

The following are examples:

Nuclear Winter

This idea from the early 1980s arose from the popularity of computer models and was promoted by the physicist Carl

---

[48] *Environmental False Alarms*, http://www.akdart.com/enviro6.html

The Global Warming Scam

Sagan.[49] It was claimed that a serious nuclear war would not only cause many deaths and a disruption of human existence but would cause so much soot and other aerosols in the atmosphere that all plant life on earth would die.

The use of nuclear weapons would undoubtedly cause considerable harm. Since some 93% of them are in the arsenals of Russia and the USA the priority is avoiding a war between them, something we hope has become increasingly remote. Why should we believe nuclear bombing of modern towns, largely built from concrete and steel, would contribute the vast amounts of smoke required to support the nuclear winter theory?

At the time this theory was put forward, I was a member of the Campaign for Nuclear Disarmament. I joined it after I saw a television interview with Clement Attlee who had been the British Prime Minister who authorised to dropping of atomic bombs on Hiroshima and Nagasaki, where he admitted he was unaware that these bombs had a problem with radioactivity.

I and my family walked to and from Aldermaston[50] in a protest campaign that publicized the true horror of these bombs. We played a part in what we all hope will be their final elimination. The nuclear war scam was a deliberate diversion from this aim.

Sagan was so convinced by his theory that he expected that the fires in the Kuwait oilfields would create a nuclear winter, but he got it wrong.

Some who think that Beijing air pollution could be disastrous are unaware that similar problems of air pollution in Britain and elsewhere have been successfully solved without a risking a nuclear winter. When I lived in Manchester in 1951, I recall a pea soup fog where I could not see an illuminated street

---

[49] *Nuclear Winter*, http://en.wikipedia.org/wiki/Nuclear_winter
[50] The UK's nuclear weapons manufacturing and development facility.

lamp when standing below it. I was a member of the Clean Air Society at the time. This Society was dissolved when a policy of clean air was adopted by most local bodies, despite occasional lapses.

Silent Spring

Silent Spring by biologist/zoologist Rachel Carson warned of the dangers DDT allegedly posed to all manner of plant, animal, and human life.[51] These threats were so great, said Carson, that on balance they more than negated whatever benefits were to be gained from using the pesticide to prevent malaria.

Carson claimed the presence of DDT and its metabolites, DDE (Dichloro-Diphenyldichloro-Ethylene) and DDD (Dichloro-Diphenyl-Dichloroethane), caused the shells of bird's eggs to become thinner, thereby leading to an increased incidence of egg breakage and embryo death. This, Carson postulated, would severely interfere with bird reproduction and ultimately would lead to a silent spring bereft of the familiar sounds of birdsongs.

She also stated that the overall rise in U.S. cancer rates between 1940 (the dawn of the DDT era) and 1960, proved DDT was a carcinogen. She predicted DDT and other pesticides would spark a cancer epidemic that would wipe out nearly 100 percent of the human population.

After seven months of hearings in 1971 by the U.S. Environment Protection Agency, which produced 125 witnesses and 9,362 pages of testimony, EPA Judge Edmund Sweeney concluded that according to the evidence:

> *DDT is not a carcinogenic hazard to man. It is not a mutagenic or teratogenic hazard to*

---

[51] Carson, Rachel, *Silent Spring*, 1962, Houghton Mifflin

> man...[and the] use of DDT under the regulations involved here do not have a deleterious effect on freshwater fish, estuarine organisms, wild birds or other wildlife.

The banning of the use of DDT has caused great harm to our precious environment because supposedly *organic* agriculture campaigns against improved genetically engineered crops which use less land. The production of *biofuels* also require extra land, which has to come from clearing native reserves, with harm to the native birds. The cost of food is increased and so world poverty.[52]

The withdrawal of DDT is thought to have caused the deaths of an extra 50 million victims of malaria.

In September 2006, the World Health Organisation (WHO) announced that it would henceforth actively support indoor spraying of the chemical as a prevention of malaria not only in epidemic areas but also in areas with constant and high malaria transmission, including throughout Africa. The scientific and programmatic evidence clearly supports this reassessment.

Dr. Anarfi Asamoa-Baah, WHO assistant director-general for HIV/AIDS, tuberculosis, and malaria, said:

> *DDT presents no health risk when used properly.*

The WHO issued a statement asserting:

> *DDT provides the most effective, cheapest, and safest means of abating and eradicating*

---

[52] *Discover the Networks*, http://www.discoverthenetworks.org/viewSubCategory.asp?id=1259

infectious diseases like malaria and typhus, which may have killed half of all the people that ever lived.

The following table lists the consequences of this decision. It shows that 18 countries are now either making use of DDT or planning to do so.[53]

---

[53] Stockholm Convention on Persistent Organic Pollutants, http://www.pops.int/documents/ddt/Global%20status%20of%20DDT%20SSC%20Oct08.pdf

Table 1. Annual global production and use of DDT (in 10^3 kg a.i.) in 2003, 2005 and 2007. "n.a." denotes data not available.

Country	2003	2005	2007	Comment	Source[a]
A. Production of DDT for vector control					
1 China[b]	450	490	n.a.	for export	Pd
2 Korea DPR	n.a.	n.a.	5	plus 155 t for use in agriculture	UNITAR
3 India[c]	4100	4250	6344	for malaria and leishmaniasis	Pd, Ws, Dc
Global production	4550	4740			
B. Use of DDT for vector control					
1 Cameroon	0	0	0	plan to pilot in 2009	WHO
2 China	0	0	n.a.	discontinued use in 2003	SC
3 Congo	0	0	0	plan for reintroduction	WHO
4 Korea, DPR	n.a.	n.a.	5	plus 155 t used in agriculture	UNITAR
5 Eritrea	13	15	15	epidemic prone areas	Qu, WHO
6 Ethiopia	272	398	371	epidemic prone areas	WHO, Ws
7 Gambia	0	0	0	use starting in 2008	WHO
8 India	4444	4253	3188	for malaria and leishmaniasis	WHO, Dc
9 Madagascar	45	0	0	plan to resume use in 2009	Qu
10 Malawi	0	0	0	plan to pilot in 2009	WHO
11 Mauritius	1	1	0	to prevent malaria introduction	Qu
12 Morocco	1	1	n.a.	for occasional outbreaks	Qu
13 Mozambique	0	308	n.a.	reintroduction in 2005	WHO
14 Myanmar	1	1	n.a.	phasing out	Ws
15 Namibia	40	40	40	long-term use	WHO
16 Papua New Guinea	n.a.	n.a.	n.a.	unknown amounts used	
17 South Africa	54	62	66	reintroduction in 2000	Qu, WHO
18 Sudan	75	n.a.	0	no recent use reported	Qu, WHO
19 Swaziland	n.a.	8	8	long-term use	WHO
20 Uganda	0	0	0	High Court prohibited use, 2008	SC, media
21 Zambia	7	26	22	reintroduction in 2000	Ws, Qu, WHO
22 Zimbabwe	0	108	12	reintroduction in 2004	WHO
Global use	4953	5219	3725		

[a] Dc: Direct communication with national authorities; Pd: Project proposals submitted to the Global Environment Facility; Qu: Questionnaire on DDT by the Secretariat of the Stockholm Convention; SC: Documents published by the Secretariat; Ws: Workshop presentations in the context by country delegates of the Stockholm Convention
[b] The figure for 2005 was extrapolated from the total production. In addition to production for vector control, DDT is produced for Dicofol manufacture (approx. 3800 t p.a.) and for antifoulant paints (approx. 200 t p.a.).
[c] In addition, DDT is produced for Dicofol manufacture (approx. 280 t p.a.)

Vincent Gray

World Dynamics and Limits to Growth

The United Nations conference on the *Human Environment*, held in Stockholm in June 1972, was dominated by the sensation that had been caused by the publication of two books, *World Dynamics*[54] by Jay W. Forrester and *Limits to Growth*[55] by Donella H. Meadows, Dennis L. Meadows, Jørgen Randers and William W. Behrens III.

Forrester was a systems-dynamics analyst and Professor of Management at the Alfred P. Sloan School of Management at the Massachusetts Institute of Technology. He devised an early computer model intended to simulate the world responses to environmental change. His parameters were population, resources, pollution, capital and agriculture. Fig 3.1 shows the multitude of feedback loops, from which he derived his World2 model.

His book actually lists all the equations he used, taking up only two pages, a modest list by today's standards. Figure 3.2 shows the main predictions of the standard run of his model, where decline is caused by depletion of resources. Steadily increasing resource costs inhibit industrial growth. This has the effect of lowering standards and causing population to decline. Pollution plays no part here, unless the resource or capital declines are inoperative. Note that the population decline is scheduled from 2020, and the quality of life declines from 1950.

---

[54] Forrester, Jay W. 1971, *World Dynamics*, Wright-Allen Press, Cambridge Mass, USA

[55] Meadows, Donella H, Meadows, Dennis L, Randers Jørgen and Behrens III William W, 1972, *Limits to Growth*, Universe Books, New York and, Earth Island, London

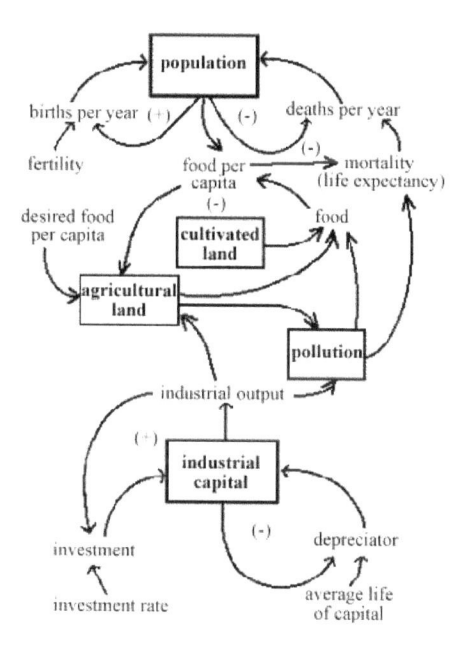

```
                    population
births per year (+)    ↕    (-) deaths per year
                      (-)
  fertility    food per  →  mortality
               capita       (life expectancy)
                 (-)
desired food         food
 per capita   cultivated
                land
           agricultural
              land
                          pollution
          industrial output →
        (+)
            industrial
             capital
                 (-)
    investment        depreciator
       ↑              ↑
  investment rate   average life
                    of capital
```

Figure 3.1
Feedback Loops

The trick is to ignore the lessons of history, which show how human progress is dependent on human ingenuity. If this disappears, disaster obviously follows.

Cole et al in their critique of *Limits to Growth* have shown how this is done.[56]

---

[56] Cole H S D, Freeman, C, Jahoda M, and Pavitt, K E R, 1973, *Thinking About the Future: A Critique of The Limits to Growth*, Chatto and Windus

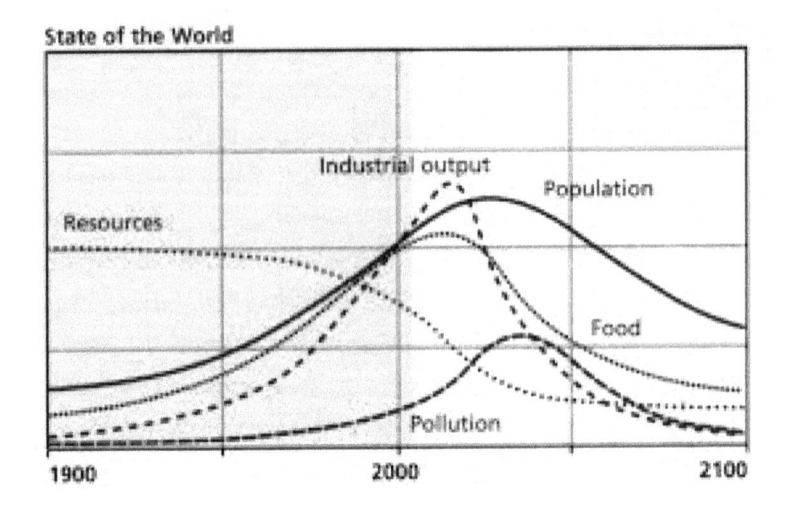

**Figure 3.2
World Forecasts**

The Population Bomb

Paul R Ehrlich, in *The Population Bomb*[57] and *The Population Explosion*[58] (with Ann H. Ehrlich), publicised the belief that the world's population is too large and is exceeding its resources.

Malthus claimed that, if unrestrained, population will increase exponentially.[59] But all organisms are restrained by the struggles of evolution with other organisms. Population of any organism depends on success in finding food, coping with predators and ingenuity in expanding territory. Human

---

[57] Ehrlich, Paul R, 1971, *The Population Bomb*, Ballantine
[58] Ehrlich, Paul R, Ehrlich, Ann H., 1990, *The Population Explosion*, Simon and Schuster
[59] Malthus, T. R., 1798, *An Essay on the Principle of Population*, Oxford World Classics

population increase is closely related to success in all of these factors. If we avoid the advice of the above authors there is no reason why this should not continue.

Much is made of the current increases in human population but little attention is paid to its lack of uniformity.

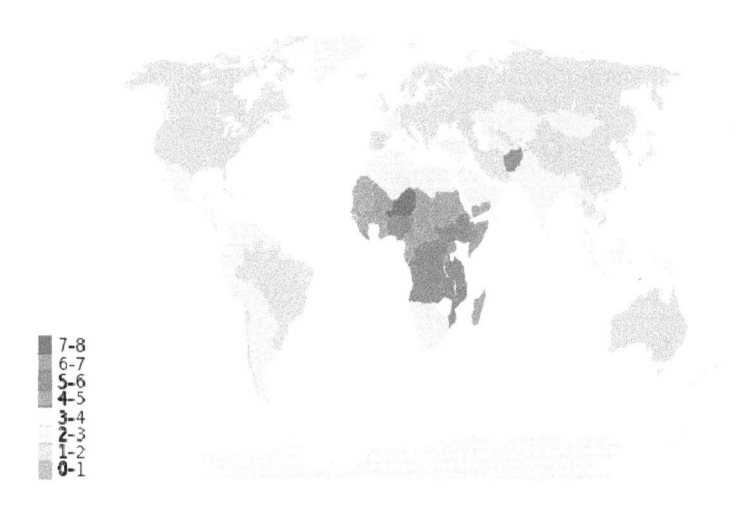

**Figure 3.3
Fertility Rates in Different Countries**

The above map shows that fertility rates (the expected number of children born per woman in her child-bearing years).[60]

The fertility rates in most developed countries are currently below the level needed to replace the present population. In undeveloped countries, particularly most of Central Africa and a part of India, fertility rates are above the ability of the local

---

[60] *Fertility Rates in Different Countries*, http://en.wikipedia.org/wiki/List_of_sovereign_states_and_dependent_territories_by_fertility_rate

economy to provide support for them.

Population programmes depend on the locality. They need to concentrate on encouragement for childbearing in advanced countries, but for Central Africa and part of India there are needs for contraceptive advice, empowerment of women and industrial development.

The United Nations population Fund has recently released a detailed forecast of population changes until 2300 which bear little relationship to the beliefs of Ehrlich and his associates.[61]

Depletion of Resources

Ehrlich and others believe the Earth's resources are becoming scarcer. Expanding population uses them at an increased rate. Therefore he predicted that the prices of these resources should increase.

In 1980, Julian L Simon offered Ehrlich a bet.[62] Ehrlich could choose any five raw materials he wanted. Simon sold Ehrlich an option to buy an amount of each raw material worth $200 in 1980 dollars. If the prices increased over the next ten years, Simon would pay Ehrlich; however, if the prices decreased over the same time period, Ehrlich would have to pay Simon.

Ehrlich chose five metals: copper, chrome, nickel, tin and tungsten. The bet was on. Ten years later, after adjusting for inflation, just as predicted, the prices of all five metals went down. Ehrlich had lost. He sent Simon a check and nothing else. Simon offered to bet again and up the ante to $20,000; Ehrlich

---

[61] UN Population Fund, *Population to 2300*, http://www.un.org/esa/population/publications/longrange2/WorldPop2300final.pdf

[62] Simon-Ehrlich Wager, http://en.wikipedia.org/wiki/Simon%E2%80%93Ehrlich_wager

declined.

Wikipedia and all the others try to argue that Simon was lucky. He might have been less successful if he had chosen a different time period or set of commodities. But he did not and succeeded in deflating Ehrlich's pretensions.

Fortunately, the real world constantly changes. What constitutes resources also changes, related to the requirements of society and its ingenuity in meeting them.

Let us take an example from energy resources, which are so often claimed to be depleted.

The following figures are from the BP Statistical Review of World Energy 2013.[63] They show how the proved resources of various forms of energy have changed.

---

[63] *BP Statistical Review of World Energy*, 2013, http://www.bp.com/content/dam/bp/pdf/statisticalreview/statistical_review_of_word_energy_2013.pdf

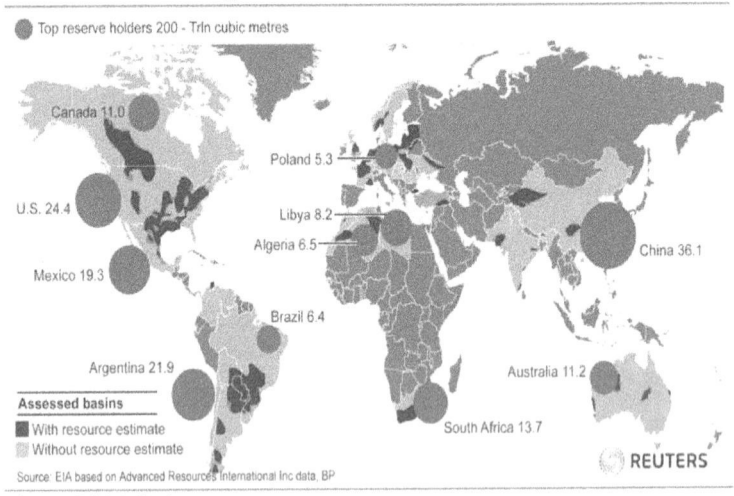

Energy	1992	2002	2011	2012
Oil Barrelsx19^9	1039.3	1321.5	1654.1	1638.9
Gas m^6	177.6	154.9	187.8	187.3
Coal tonnes^6	981.8	984.5	861.0	
Nuclear mt oil eq	610.5		586.4	
Hydro mt oil eq	598.5			851.1

Figure 3.4
World Fracking Sites

Anti Fracking

Environmentalists are modern Luddites. Any change in manufacture or technology can be construed to harm the environment, particularly extractive industries. Coal and oil should never have been discovered.

Fracking is a technique for recovering oil and gas from oil-bearing strata by its controlled damage. It has recently been improved by pumping water containing small plastic balls which can hold fissures open. It has been embraced by the United States on a grand scale, despite the supposed environmentalist support. The USA has suddenly changed from an impoverished nation dependent on oil from the Middle East to the largest oil and gas producer in the world.

Figure 3.4 shows Fracking sites around the world.[64]

Environmentalists in other nations have largely succeeded in preventing this path to prosperity in many countries. The result has been an enormous difference in price for natural gas, greatly to the disadvantage of Europe.[65]

---

[64] *Triple Pundit*, http://www.triplepundit.com/2012/06/fracking-boom-bust/

[65] *Financial Sense*, http://www.financialsense.com/contributors/gail-tverberg/why-high-oil-prices-are-now-affecting-europe-more-than-the-u-s

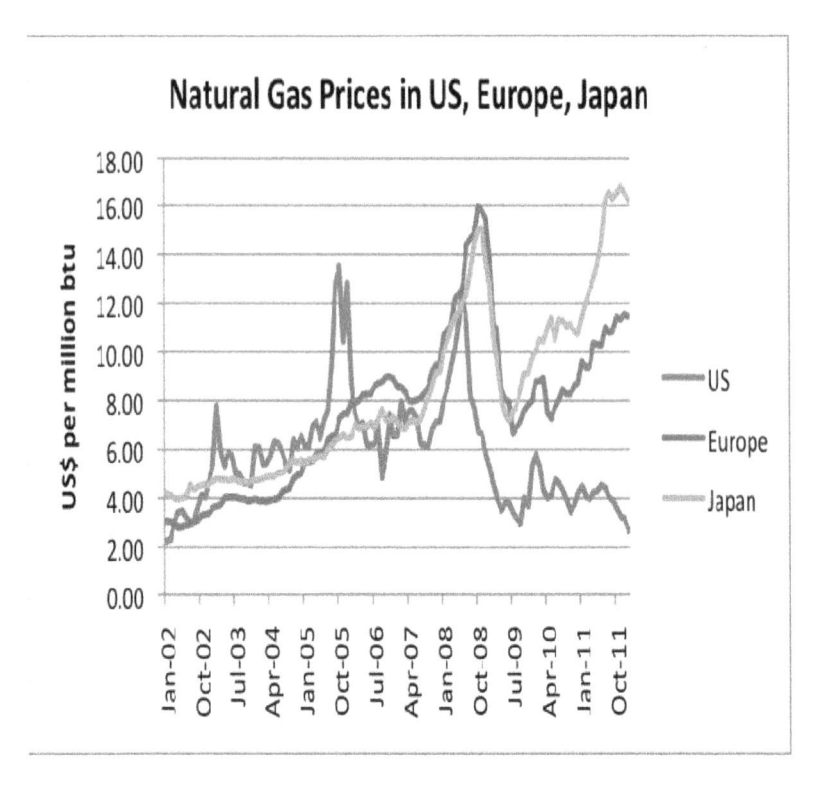

**Figure 3.5
Comparative Gas Prices**

Genetic Engineering

Genetic engineering is the process whereby we select plant varieties to provide food and utility. It has been the work of all generations and it has been the main reason why human beings increased in numbers and prosperity. Of course, our forefathers did not know the mechanisms which enabled their success. They knew nothing of genes or the mechanism of heredity. It is only after the discovery of the nature of genes and the chemistry of heredity that it has been possible to carry out the procedures of improving plant utility by using scientific knowledge.

But again, environmentalists are against it and are therefore preventing improvements in agriculture on the grounds that it damages the environment or is unsafe. Yet these limitations were always present and did not prevent our forefathers from providing the food we have today. Potential problems always existed and we should be grateful that plant and animal experiments in the past were not forbidden. Today we have the advantage of background knowledge.

The use of modern techniques for modifying genes instead of the dependence on selection of favourable varieties is said to be artificial. Selection without background knowledge is supposedly natural and therefore in some way superior.

In 2013, GM crops were planted in 27 countries; 19 were developing countries and 8 were developed countries.[66] 2013 was the second year in which developing countries grew a majority (54%) of the total GM harvest. 18 million farmers grew GM crops; around 90% were small-holding farmers in developing countries. The supposed defects of the technique have never become evident.

---

[66] Wikipedia,
http://en.wikipedia.org/wiki/Genetically modified_crops
NAPAP

Country	2013–GM planted area (million hectares)	Biotech crops
USA	70.1	Maize, Soybean, Cotton, Canola, Sugarbeet, Alfalfa, Papaya, Squash
Brazil	40.3	Soybean, Maize, Cotton
Argentina	24.4	Soybean, Maize, Cotton
India	11.0	Cotton
Canada	10.8	Canola, Maize, Soybean, Sugarbeet
Total	175.2	----

Acid Rain

Acid rain is simply rain with a lower than normal pH caused by emissions of sulphur dioxide (SO_2) and nitric oxides (NO_x), typically from factories.

In 1989, the U.S. Congress passed the National Acidic Precipitation Assessment Program (NAPAP) with a recent Report which shows the progress that has been made in reducing these acidic gases.[67]

---

[67] http://ny.water.usgs.gov/projects/NAPAP/NAPAP_2011_Report_508_Compliant.pdf.pdf

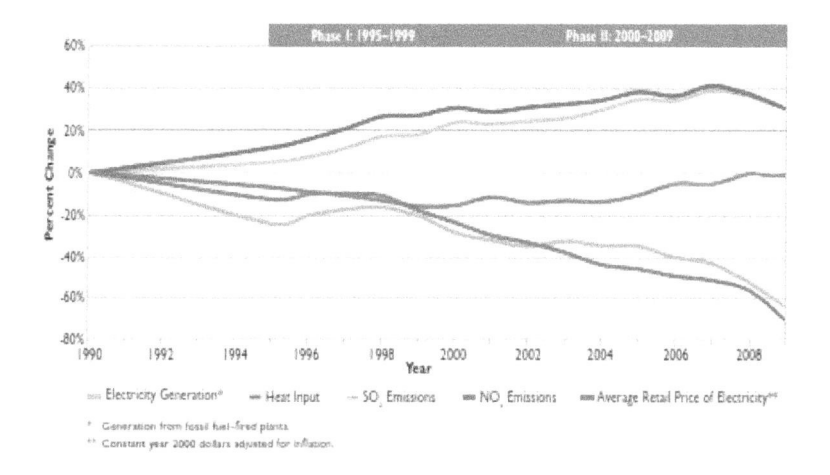

Figure 3.6
U. S. Progress in Reducing Acid Gas

Water Shortage

All organisms need water. Heat from the sun evaporates it, mainly from the oceans, and deposits it as rain, hail, snow, dew and frost. Since most precipitation on land originated from the oceans, a steady bonus is supplied by the climate to the land. Some fill or replenish natural aquifers and the surplus flows from rivers back to the sea.

Potable water is needed for drinking and cooking. Water is also needed for crops, cleaning and sewage processing. Human settlements have traditionally been chosen with local accessibility to water in mind. Human expansion has needed measures to ensure reliability of supply. The Romans pioneered the construction of reservoirs pipelines drainage and aqueducts.

Today their example pervades the world, and there is the addition of the possibility of desalinating sea water, now practised by 25 different countries.

The availability of water is usually considered to be a human right that should be available for free, but supply costs money, which has to come from somewhere to an extent that depends on location and time of year. All claims of water shortage can be met by the availability of finance, but this truth is often ignored.

Many current problems arise from reluctance to maintain, modernize or expand pipelines, reservoirs or water treatment plants to meet increases in demand.[68] Campaigns to conserve water or limit its use are easier to do than find or spend the money to increase services. Some recent floods in Britain were caused by neglect of drainage, as recommended by environmentalists, supposedly to benefit *endangered species.*

---

[68] *Water Scarcity*, http://en.wikipedia.org/wiki/Water_scarcity

The Ozone Hole

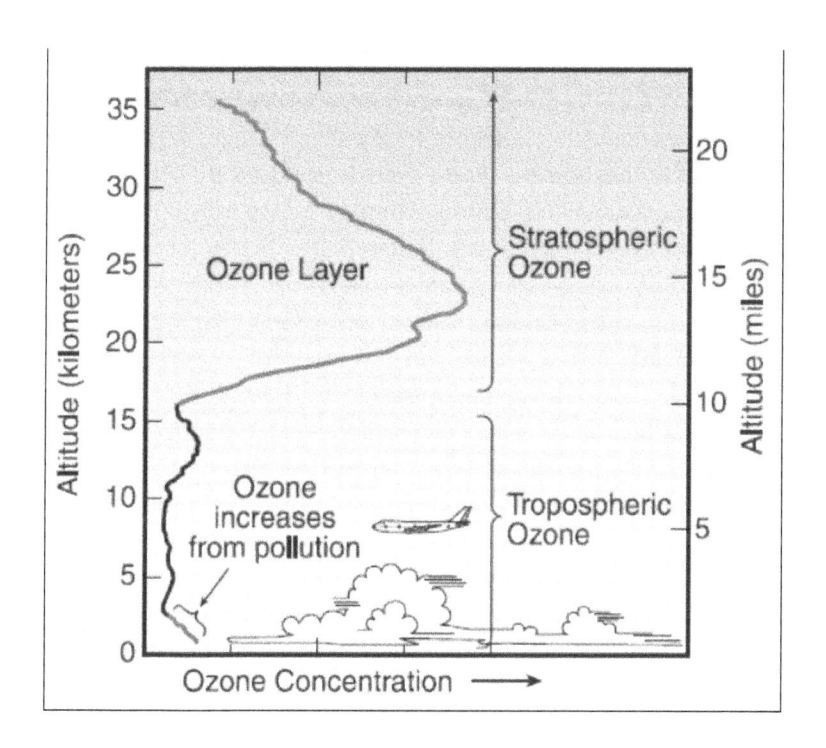

**Figure 3.7
Ozone in the Atmosphere**

Ozone is a form of oxygen with three atoms to a molecule which has a higher concentration in the stratosphere through the action of the ultraviolet part of sunlight.[69] In this process, it reduces the amount of ultraviolet light coming to the earth, particularly the

---

[69] *The Ozone Hole*, http://www.theozonehole.com/

short wavelength region, UVB radiation (290-320nm) as opposed to UVA radiation (320-400nm).

Gordon Dobson at the University of Oxford in the 1920s built the first instrument to measure total ozone from the ground, now called the *Dobson Ozone Spectrophotometer*. In tribute, ozone is measured in Dobson Units. The area of the ozone hole is determined from a map of total column ozone. It is calculated from the area on the Earth that is enclosed by a line with a constant value of 220 Dobson Units.

In 1984, a recurring Antarctic reduction of ozone was noticed in spring and summer which recovered as the polar vortex arrived in autumn. It was referred to as the *Ozone Hole*.

In 1974, Rowland and Molina did experiments that suggested the ozone hole could be caused by chlorinated hydrocarbons emitted in the atmosphere.[70] It was also suggested that the effect would damage human health as it would increase human exposure to the shorter-wavelength ultra violet radiation (UVB).

This suggestion was widely adopted and led to the Montreal Protocol on Substances that Deplete the Ozone Layer, agreed 16 September 1987 at the Headquarters of the International Civil Aviation Organization in Montreal. This protocol has subsequently endorsed by almost all nations.

The Montreal Protocol stipulates that the production and consumption of compounds that are claimed to deplete ozone in the stratosphere—chlorofluorocarbons (CFCs), halons, carbon tetrachloride and methyl chloroform—are to be phased out by 2000 (2005 for methyl chloroform).

This action virtually eliminated a prosperous and useful industry with applications such as refrigeration, dry cleaning, anaesthetics and spray cans and its replacement largely with

---

[70] Molina, M. J. and Rowland, F. S., 1974, *Stratospheric Sink for Chlorofluoromethanes*, Nature, 249 810-812

inflammable hydrocarbons.

Maduro and Scahauerhammer[71] in 1992 and Dixy Lee Ray[72] in 1993 published criticisms of this procedure. Used copies of both books can currently be purchased from Amazon.com for $0.01, so they are currently little regarded.

These books cast doubt on the decision to ban chlorinated hydrocarbons. They point out the lack of information about the stratosphere, its variability and chemistry, which has largely been studied with laboratory experiments. There are several other sources of chlorine. The ozone concentration may naturally vary.

Singer recently showed that doubts still apply.[73] In his review of recent literature, he shows that the evidence is doubtful and the potential damage is unproven. It is even suggested that the supposedly damaging UVB radiation reacts with ozone in the troposphere so it may never arrive on the surface.

Ever since the Protocol was agreed to in 1987, we have been told how the ozone hole is being eliminated as a result of this action. Yet this is untrue.

Figure 3.8 gives the latest figures from the NOAA for 2002-2013.[74] It shows that the size of the hole fluctuates, but shows no signs of disappearing.

---

[71] Maduro, R. A., and Schauerhammer, 1992, *The Hole in the Ozone Scare 21st Century*, Science Associates

[72] Ray, Dixy Lee and Guzzo, L., 1993, *Environmental Overkill*, Regnery Gateway, Washington DC, USA

[73] Singer F., 2010, *Hasty Action, Shaky Science*, http://heartland.org/policy-documents/ozone-cfc-debacle-hasty-action-shaky-science

[74] NOAA Stratospheric Ozone, http://www.cpc.ncep.noaa.gov/products/stratosphere/polar/gif_fil es/ozone_hole_plot.png

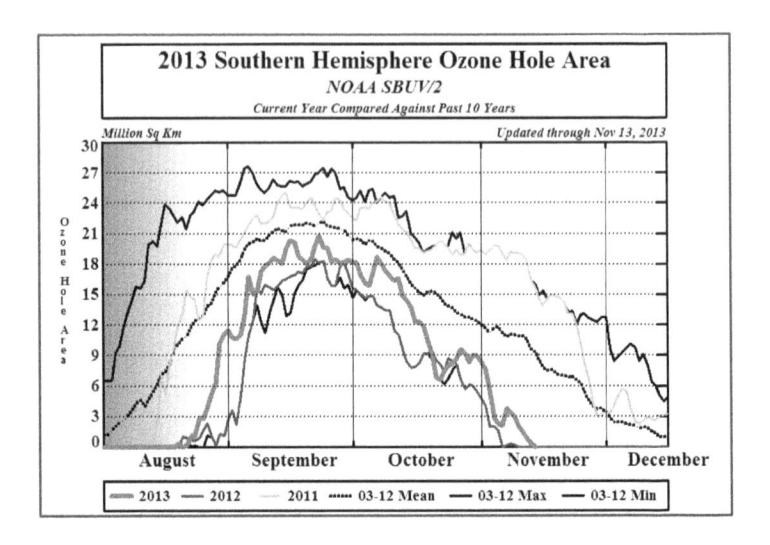

**Figure 3.8
The Southern Hemisphere 2003-2013**

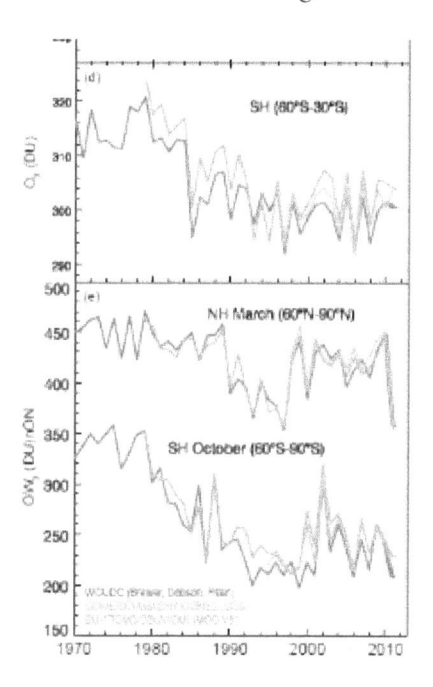

**Figure 3.9
Reduction of the Ozone Hole**

Figure 3.9 is from the latest IPCC Report which gives figures from 1970.[75]

The ozone hole has not gone away and there is no evidence that anybody has suffered harm from its existence.

---

[75] IPCC, 2013, *Climate Change 2013: The Physical Science Basis. Contribution of Working Group I to the Fifth Assessment Report of the Intergovernmental Panel on Climate Change*, Stocker, T.F., Qin, D., Plattner, G-K, Tignor, M., Allen, S.K., Boschung, J., Nauels, A., Xia, Y., Bex, V., and Midgley, P.M., editors, Cambridge University Press, Cambridge, United Kingdom and New York, NY, USA, 1535 p. 27

Vincent Gray

Further environmental scams are given by Simon.[76]

Global Warming Scam and Climate Change Superscam

There is no question that the most one of the most successful scams ever perpetrated is by the environmental movement in the claim that the climate is controlled by human emissions of carbon dioxide and other minor trace gases.

---

[76] Simon, Julian L, 1998, *The Ultimate Resource2,* Princeton University Press,
http://www.forecastingprinciples.com/index.php?option=com_content&view=article&id=340&Itemid=108 el on Climate Changehttp://www.juliansimon.com/writings/Ultimate_Resource/

The Global Warming Scam

CHAPTER 4: THE IPCC SUPERSCAM

CHAPTERS 2 AND 3 shows how the environmental heresy has sought to justify the pseudo-religious dogma that humans are destroying the earth by launching scams intended to justify this belief. Then they decided to concentrate on the theory that humans are damaging the climate by emissions of so-called greenhouse gases. This claim has been so successful that it deserves the title of *superscam*.

After the publication of several books about the alleged threats to the planet and a 1970 Study of Environmental Problems (SCEP), there arose a shopping list of man's impact on the global environment with the preparation of a short list of scams that could be exploited further.[77] They then chose as the prime candidate the theory that emissions of greenhouse gases were destroying the earth by causing global warming.

From these beginnings, a number of local meetings about the environment led to the first United Nations Conference on

---

[77] *Report of Study of Critical Environmental Problem*, https://mitpress.mit.edu/authors/study-critical-environmental-problems-scep

the Environment in Stockholm in 1972.[78] It established the United Nations Environment Programme (UNEP) and issued a long Declaration which embodied the main environmentalist dogmas and demands.

The first World Climate Conference, organised by the World Meteorological Organisation (WMO) in 1979, called on all nations to unite in efforts to understand climate change and to plan for it, but it did not call for action to prevent future climate change.

A subsequent conference in Villach, Austria in 1980, organised jointly by the United Nations Environment Programme (UNEP), World Meteorological Organisation and the International Council for Science (ICSU), also concluded that the potential threats were sufficient to warrant an international programme of cooperation in research but that, due to scientific uncertainties, the development of a management plan for CO_2 would be premature. A follow-up assessment was programmed for 1985, again to take place in Villach.

The Villach conference of October 1985 is widely credited with being critical to the placing of the climate change swindle firmly on the international political agenda, and to the subsequent establishment of the Intergovernmental Panel on Climate Change (IPCC).

The conference concluded with this scare:

> ...as a result of the increasing greenhouse gases it is now believed that in the first half of the next century (21st century) a rise of global mean temperature could occur which is greater than in any man's history.[79]

---

[78] Spiked, http://www.spiked-online.com/newsite/article/3540#.VMQWyC5JfEQ
[79] Ibid.

The Global Warming Scam

The conference statement maintained that:

> *While some warming of climate now appears inevitable due to past actions, the rate and degree of future warming could be profoundly affected by government policies on energy conservation, use of fossil fuels, and the emission of greenhouse gases.*

In 1987, the 10th Congress of the World Meteorological Organisation recognized the need for objective, balanced, and internationally coordinated scientific assessment of the understanding of the effects of increasing concentrations of greenhouse gases on the Earth's climate and on ways in which these changes may impact socio-economic patterns.

In its follow up, the WMO Executive Council asked the Secretary General of WMO, in coordination with the Executive Director of The United Nations Environment Programme (UNEP), to establish an ad hoc intergovernmental mechanism to provide scientific assessments of climate change. Thus the Intergovernmental Panel on Climate Change (IPCC) was formed in 1988.

The publishing of the IPCC's first assessment report in 1990[80] provided the basis for the 1992 United Nations Conference on Environment and Development—also known as the *Earth Summit*—in Rio de Janeiro, which led to the United Nations Framework Convention on Climate Change (UNFCCC)

---

[80] Houghton, J. T., Jenkins, G. J., and Ephraums, J.J., editors, 1990, *Climate Change : The IPCC Scientific Assessment*, Cambridge University Press, Preface, pages xxi, 202

which was ready for signature at the Conference.[81]

The Intergovernmental Panel on Climate Change (IPCC) was set up on order to:

> *Assess available scientific information on climate change: Working Group I.*
> *Assess the environmental and socio-economic impacts of climate change: Working Group II,*
> *Formulate response strategies: Working Group III*

The second and third objectives depend heavily on the first. This book deals mainly with the reports issued by the Working Group I since its opinions form the background for the activities of Working Groups II and III.

Later, Reports further defined the Working Groups as follows:

> *The IPCC Working Group I WGI deals with the Science of Climate Change.*
> *Working Group II WGII with Impacts, Adaptation and Vulnerability.*
> *Working Group III, WGIII with Mitigation of Climate Change.*

The three Working Groups are made up of nominees of the two sponsors and of the Governments that support the greenhouse delusion. The scientists are mainly Government employees, or recipients of Government finance. As Governments throughout the world have come to adopt policies dependent on the belief

---

[81] *United Nations Framework Convention on Climate Change*, http://unfccc.int/essential_background/convention/background/items/1349.php

that greenhouse gas emissions are causing harmful effects on the climate, all of the Working Group members tend to be supporters of this view as are the "Lead Authors" of the Reports who are nominated by them.

Drafts of all the main Reports of the IPCC are circulated for comment. Initially this was only to Government Environment Departments, who then consulted with local experts and interested parties before forwarding comments received. Nowadays almost anyone can comment, provided they tell the right story to the IPCC. There are three drafts of each Report, the third being circulated only to Governments. There is evidence that some of the most extravagant claims only appear in the Final Draft without public review.

The Framework Convention on Climate Change

The 1992 Framework Convention on Climate Change (FCCC), adopted on May 9th, 1992 came into force on 21st March 1994.[82] By that time there were 166 signatures from National Governments.

This Convention initiated a system for compulsory reduction of greenhouse gases by *Annex I* Governments, which has become progressive. The provisions have not been entirely enforced, but they are still causing major economic disaster in many countries.

The procedure has been implemented by a series of *Conferences of the Parties* (COP 1,2,3, etc) in the different major cities of the world, including subsidiary meetings for implementation of the other campaigns of the environmental movement. At this writing, these meetings have reached to COP21 which will take place in Paris in December 2015 where

---

[82] Ibid.

attempts are being planned to impose global restrictions or disastrous proportions.

The IPCC Reports are a major contribution to the progress of the increasing restrictions on economic activity resulting from the main COP meetings, and their reports have all been prepared in order to influence the successive meetings.

The Climate Change Superscam

The FCCC defined *Climate Change* in Article 1 as follows:

> ...a change of climate which is attributed directly or indirectly to human activity that alters the composition of the global atmosphere and which is in addition to natural climate variability observed over comparable time periods.[83]

It was as if they had followed the advice of Humpty Dumpty in *Alice through the Looking Glass*:

> "When I use a word," Humpty Dumpty said, in rather a scornful tone, "it means just what I choose it to mean—neither more nor less."

The FCCC statement is legally binding on the Governments who signed the Convention. It amounts to an assertion that all change in the climate is caused by human emissions of greenhouse gases, even when it is only "attributed, directly or indirectly." There is no need for evidence, scientific or otherwise. As they define it, *change of climate* that is *natural* is mere *variability*.

This legally binding definition provides a license for the

---

[83] Ibid.

wholesale distortion of climate science carried out by the IPCC in its many publications.

They admit that human activity that alters the composition of the global atmosphere is only ONE possible factor in our changing climate—there is also natural climate variability which contributes to a changing climate.

They play the trick of suggesting that natural changes in the climate are not changes at all, but are merely *variables*. This permits them to assume that any change which is *unprecedented*, because it cannot be shown to be variable, must be caused by their greenhouse gases.

Climate Change, which is the alteration of the composition of the global activity by human activity, is a kind of religious slogan and as such it is an article of faith. No evidence or proof is needed; indeed it is better absent as this is a test of religious zeal.

Proof is, anyway, unnecessary. All that is needed is for the proposition to be *attributed directly*, or even *indirectly*. It is true even by an *attribution* by one single lunatic.

But to put a scam like this over on the public there must be *attribution,* not only by the devotees of the environmentalist faith whose support is guaranteed, but also by a collection of pundits who could influence politicians. You must line up celebrities, film stars, sports heroes and scientists. Better still, set up political parties devoted to the promotion of this scam.

The Intergovernmental Panel on Climate Change was set up to pay scientists to support the scam. Science jobs were scarce, but here, money was no object. A profitable career was offered with foreign travel, guaranteed publications supported by reliable editors and peer reviewers and even a possible Nobel Prize.

The only condition was that they would all *attribute* almost everything about the climate to *human activity that alters the composition of the global atmosphere* to a level satisfactory to the

Government Representatives that controlled the IPCC and the signatories of the FCCC on pain of dismissal and loss of career if they failed to do so. A suitable system of justification for this process was built up using the latest principles of public relations and political and verbal spin.

The FCCC definition was supported by the First Report and in the Supplementary Reports, 1992[84] and 1994[85], but the 1995[86] and subsequent reports had, as a footnote on the first page, a different definition of *climate change*.

This reads as follows in the latest, 2013 Report:

Climate Change

> *Climate change refers to a change in the state of the climate that can be identified (e.g., by using statistical tests) by changes in the mean and/or the variability of its properties, and that persists for an extended period, typically decades or longer. Climate change may be due to natural internal processes or external forcings such as modulations of the solar cycles, volcanic eruptions and persistent anthropogenic changes*

---

[84] Houghton, J. T., Callander, B. A., and Varney, S. K., editors, 1992, *Climate Change 1992: The Supplementary Report to the IPCC Scientific Assessment*, Cambridge University Press

[85] Houghton, J. T., Meira Filho, L. G., Bruce, J., Hoesung Lee, Callander, B. A., Haites, E., Harris, N. and Maskell, K., editors, 1995, *Climate Change 1994: Radiative Forcing of Climate Change and an Evaluation of the IPCC IS92 Emission Scenarios*, Cambridge University Press

[86] Houghton, J. T., Meira Filho, L. G., Bruce, J., Hoesung Lee, Callander, B. A., Haites, E., Harris, N. and Maskell, K., editors, 1996, *Climate Change 1995: The Science of Climate Change*, Cambridge University Press, p. 4

> *in the composition of the atmosphere or in land use. Note that the Framework Convention on Climate Change (UNFCCC), in its Article 1, defines climate change as: 'a change of climate which is attributed directly or indirectly to Human activity, that alters the composition of the global atmosphere and which is in addition to natural climate variability observed over comparable time periods.'* [87]

The UNFCCC makes a distinction between climate change attributable to human activities altering the atmospheric composition and climate variability attributable to natural causes—which it claims are variable and so may be ignored.

The IPCC make a similar distinction. There is the one the term REFERS to and the one which does not. *Climate change* REFERS to changes that can be identified by statistical tests which show they are PERSISTENT for an extended period, typically decades or longer On the other hand there is *climate change* which is not being REFERRED TO which is not PERSISTENT for decades or longer which includes external forces such as solar changes (which they think, wrongly, are cyclic), volcanic eruptions, and human induced changes in land use.

They seem to think that the persistent changes will turn out

---

[87] IPCC, 2013, *Climate Change 2013: The Physical Science Basis*, Contribution of Working Group I to the Fifth Assessment Report of the Intergovernmental Panel on Climate Change, Stocker, T.F., Qin, D., Plattner, G-K, Tignor, M., Allen, S.K., Boschung, J., Nauels, A., Xia, Y., Bex, V., and Midgley, P.M., editors, Cambridge University Press, Cambridge, United Kingdom and New York, NY, USA, 1535 p. Glossary

sufficiently representative, standardized and symmetrical, to be capable of statistical treatment that can give a satisfactory mean and estimate of its variability.

The chosen climate change which is supposed to be persistent has turned out to be less persistent that the climate change which is not chosen. They have supplied no less than 63 possible reasons why this is so.[88]

The only difference between the IPCC definition of *climate change* and that of the FCCC is that they have added anthropogenic changes in land use to *natural variability* as that part of change of climate which can be considered to be merely variable. This assumption is immediately dubious.

For the models, the less *persistent, natural* effects are considered constant. They even try to argue that such natural effects are themselves caused by humans.

Their latest definition of Climate is as follows is as follows:

Climate

> *Climate in a narrow sense is usually defined as the average weather, or more rigorously, as the statistical description in terms of the mean and variability of relevant quantities over a period of time ranging from months to thousands or millions of years.*
>
> *The classical period for averaging these variables is 30 years, as defined by the World Meteorological Organization. The Relevant quantities are most often surface variables such as temperature, precipitation and wind.*

---

[88] The Hockey Schtick,
http://hockeyschtick.blogspot.co.nz/2014/07/updated-list-of-29-excuses-for-18-year.html

> *Climate in a wider sense is the state, including a statistical description, of the climate system.*[89]

Nowhere in any part of the IPCC Reports are there any actual figures for *statistical description in terms of the mean and variability of relevant quantities* for any locality, even for temperature, precipitation or wind direction or for the recommended period of 30 years.

Since every local climate is different, global climate consists of an amalgamation of every local climate, an attempt to provide means and variability globally is impossible.

Their claim that they are able to provide a plausible *statistical description of the climate system* is simply false.

The 2013 IPCC definition of the Climate System is:

Climate System

> *The climate system is the highly complex system consisting of five major components: the atmosphere, the hydrosphere, the cryosphere, the lithosphere and the biosphere, and the interactions between them. The climate system evolves in time under the influence of its own internal dynamics and because of external forcings such as volcanic eruptions, solar variations and anthropogenic forcings such as the changing composition of the atmosphere and land use change.*
>
> *The **atmosphere** is the gaseous envelope surrounding the earth*

---

[89] Ibid.

> *The **hydrosphere** is all the surface water*
> *The **cryosphere** is all surface and below surface regions of solid ice*
> *The **lithosphere** is the upper layer of solid earth, both land and water, but not volcanoes*
> *The **biosphere** is all living organisms both land and sea.*[90]

There is no mention of the reason why the climate exists: the sun.

As explained in Chapter 1, the climate is a heat engine which has the function of supplying the energy from the sun to maintain the living organisms on the earth. In the process, there is a complex dynamic interaction between a large collection of measurable climate properties. The division of these properties into intellectual disciplines fails to integrate them to their prime function, the provision of energy for life.

By doing this, they concealed the reality of the climate. In order to place emphasis on only one of its many features and try to persuade us that it is of overwhelming importance, they postulated a dead, static earth where there is no provision for human life except change of land use and the emission of greenhouse gases.

This strange static picture of a *pseudoclimate* with a flat earth, constant sunshine, no wind and vertical radiation is shown in IPCC diagrams such as Figure 4.1: [91]

---

[90] Ibid.
[91] Solomon, S., Qin, D., Manning, M. R., Marquis, M., Averyt, K., Tignor, M. H., Miller, H. L., and Chin, Z., editors, *Climate Change 2007: The Physical Science Basis*, IPCC, C13 11

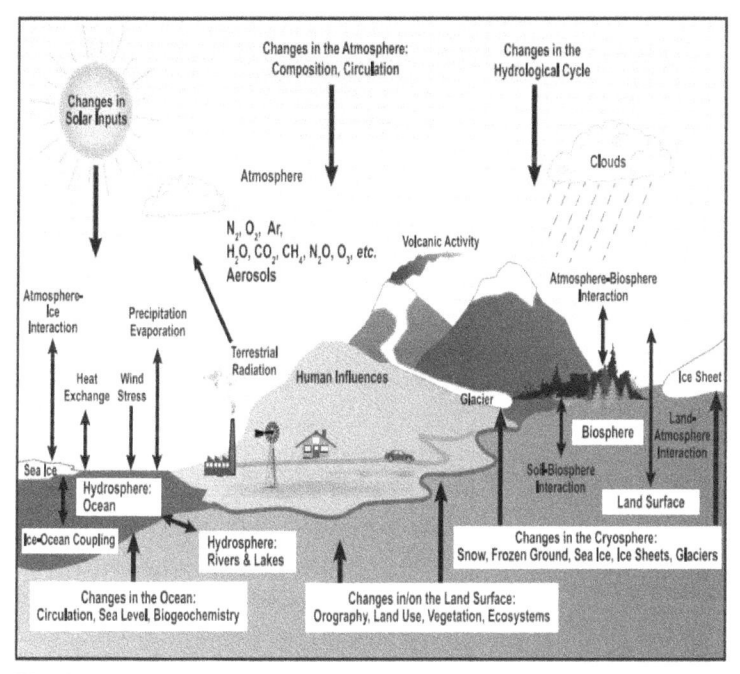

Figure 4.1
Idealized Flat Earth Climate

The IPCC Reports

The IPCC exercise was set up in order to accumulate evidence that the globe is undergoing global warming as a result of increases in carbon dioxide and other greenhouse gases in the atmosphere. There was never any intention to provide a balanced or unbiased scientific assessment of climate science.

From the beginning, there were scientists who disagreed

with the theory that increases in greenhouse gases are harmful, but their views have not been included in the IPCC Reports and comments made by them have been comprehensively rejected, to the extent that few now bother to comment at all. Some recognised experts have resigned or expressed their opposition to the entire exercise.

This deliberate bias was made clear in Appendix 4 of the first Working Group I IPCC Report, 1990, in an introduction to a list of Reviewers, with the statement:

> *While every attempt was made by the Lead Authors to incorporate their comments, in some cases these formed a minority opinion which could not be reconciled with the larger consensus.*

The *Summary for Policymakers* arises because the Governments that sponsored the Report wish to authorize it and ensure that it corresponds with their *Climate Change* policies. It is agreed line-by-line by anonymous Representatives of the Governments. It is drafted mainly by selected scientists from the main Report. It is sometimes not understood that they are acting on orders, not as independent scientists.

The *Summary for Policymakers* is actually a *Summary BY Policymakers* as it is not just advice to other policymakers, it is a summary carried out and approved by the policymakers themselves. It is also a genuine consensus of their views, agreed by all of them, and it does not necessarily coincide with the views of any Government or of the scientists who participate in the Report.

The Government Representatives who control the Reports are never named. In addition to their influence on the Summary for Policymakers, they play a large part in selecting and approving the Lead Authors of the various chapters, and permitting the views of reviewers to be considered.

The Chapters of each Report are arranged in such a way as to promote the idea of *climate change* is caused by *greenhouse gas* increases. Actual climate observations are *obscured, smoothed, filtered, linearized, interpolated* and converted into *data* with *outliers or noise* eliminated, in order to find *trends* which can be held to support the greenhouse theory.

The First Reports incorporated the findings of all three Working Groups into one volume. Subsequent Reports issued separate volumes for the proceedings of WGII and WGIII and from the third one there was in addition a Synthesis Volume which tried to summarise all three. This Chapter deals in detail only with the WGI Science Reports, since the other Working Groups merely apply the results of the WGI Report.

Climate Change: The IPCC Scientific Assessment 1990

This first report, issued in 1990, was used as a background to the 1992 "Earth Summit" at Rio de Janeiro in 1992.

The Preface was signed by:

> *G.O.P. Obasi, Secretary General, World Meteorological Organisation*
> *M. K. Tolba, Executive Director United Nations Environment Programme*
>
> *Foreword, signed by Dr John Houghton, Chairman IPCC Working Group I*
> *Introduction.*
> *The Chapter Headings were:*
> *Report Prepared for IPCC by Working Group I.*
> *Edited by J. T. Houghton, G. J. Jenkins, and J. J. Ephraums. Published by the Cambridge*

Vincent Gray

University Press 1990.
Policymakers Summary prepared by Working Group I
Contents
Executive Summary
Annex: Emissions Scenarios from Working Group III
Introduction
1. Greenhouse Gases and Aerosols. R T Watson, H Rohde, H Oeschiger and U Siegenthaler
2. Radiative Forcing of the Climate K P Shine, R G Derwent, D J Wuebbles and J J Morcrette
3. Processes and Modelling U Cubasch and R D Cess
4. Validation of Climate Models W L Gates, P R Rowntree and Q Z Zeng
5. Equilibrium Climate Change J F B Mitchell, S Manabe, T Tokioka and V Meleshko
6. Time-Dependent Greenhouse-Gas-Induced Climate Change F P Breterton, K Bryan and I D Woods
7. Observed Climate Variations and Change C K Folland, K Bryan and J D Woods
8. Detection of the Greenhouse Effect in the Observations T M L Wigley and T P Barnett
9 Sea Level Rise R A Warrick and H Oerlemans
10. Effects on Ecosystems J M Melillo T V Callaghan, F J Woodward E Sa'atiand S K Sinha
11. Narrowing the Uncertainties G M Bean and J McCarthy
Annex: Climatic consequences of emissions
Appendix 1: Emissions Scenarios
Appendix 2 Organisation of the IPCC and

The Global Warming Scam

Working Groups
Appendix 3 Contributors (306 including some duplicates).
Appendix 4 Reviewers (241, with duplicates).
Appendix 5 Acronyms Institutions
Appendix 6 Acronyms Programmes and Miscellaneous
Appendix 7 Units

The Policymakers Summary is merely Prepared by IPCC Working Group I for approval by the Governments.

There is no Index. Topics are difficult to find as they are often treated in more than one Chapter. Their inherent bias is made plain from the start with the statement at the beginning of Appendix 1 as quoted above.

While every attempt was made by the Lead Authors to incorporate their comments, in some cases these formed a minority opinion which could not be reconciled with the larger consensus.

As with all the Reports, much emphasis was placed on the *Mean Annual Global Surface Temperature Anomaly Record,* which is based on scientifically unacceptable basic data (unrepresentative samples) and unacceptable average daily temperatures (based on a varying mean of maximum and minimum) from sites almost never monitored for suitability.

The Report said:

> *The size of this warming is broadly consistent with the predictions of climate models, but is also of the same magnitude as natural climate variability.*

103

Actually, the predictions of climate models are broadly inconsistent with the prediction of climate models, as will be shown in *Chapter 7: Observed Variations and Change*.

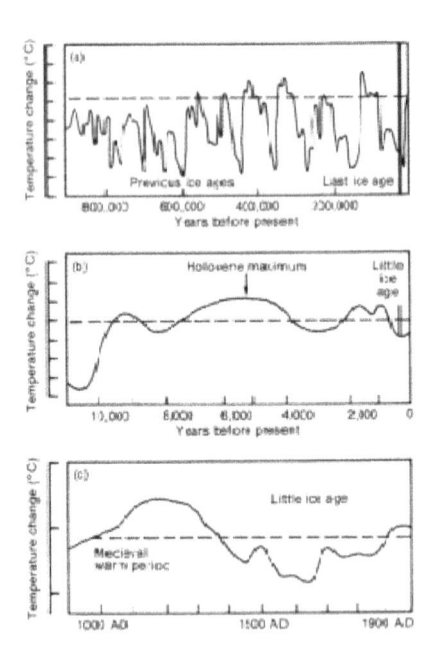

Figure 4.2
Surface Temperatures on Several Geological Scales

They established the pattern they have followed throughout of qualitative, ambiguous statements without scientific support which are invariably regarded as certain proof by their sponsors.

This first Report gave the following set of graphs as of past global temperatures which show no evidence of a relationship with atmospheric trace gases. It included the *Medieval Warm Period* and the *Little Ice Age*. These were denied in subsequent

Reports.

They blamed the trace gases for the temperature rise shown in their record from 1910 to 1940.

The Report also launched the "scenarios" of the future which are exaggerated beliefs of changes in human activity for the forthcoming century. This was the birth of the *Business as Usual* scenario. Three other sets of *scenarios* have been added since then.

The details of the scenarios were kept away from the scientists by confining the work to a sub-Committee of WGIII where they could even ignore the views of reputable economists. The scientists have found themselves encumbered with scenarios they are unable to question in the WGI Science Reports.

This Report is the only one that makes *predictions*. However, the words *We predict* occur only in the *Executive Summary* page xii right at the beginning after the Contents page. On it they *predict* a temperature rise of 0.3°C±0.2°C-0.5°C per decade, which they say is *greater than that seen over the past 10,000 years*. This hardly seems frightening and it is not supported anywhere else in the Report.

All the subsequent Reports never make *predictions*, they talk only about *projections*.

The 1992 Supplementary Report

Climate Change 1992: The Supplementary Report to the IPCC Scientific Assessment was compiled specifically to provide evidence to influence signatories for the Framework Convention on Climate Change after its adoption in May 1992.

It contains the following Chapters:

> Houghton, J T, B A Callander & S K Varney (Editors)

Foreword
Preface 1992 Supplement
A Greenhouse Gases A1 Sources and Sinks
A2 Radiative Forcing of the Climate
A3 Emissions Scenarios for IPCC: an Update
B Climate Modelling, Climate Prediction and Model Validation
C Observed Climate Variability and Change
Annex: Climatic Consequences of emissions and comparison of IS92a and SA90
Appendix 1 Organisation of IPCC and Working Group I
Appendix 2 Contributors to the IPCC WGI Report Supplement
Appendix 3 Reviewers of the IPCC WGI Report .Supplement (my comments were included as coming from R S Whitney)
Appendix 4 Acronyms
Appendix 5 Units
Appendix 6 Chemical Symbols

This time there are multiple *Authors* and also *Contributors*.

Again there was no Index.

This Report repeated the procedure of the first Report in placing *Observed Climate Variability and Change* right at the end, so that readers will not notice that observations do not agree with the models.

The Report expanded the topic of aerosols. The climate models reported in the First Report gave grossly exaggerated predictions of current temperatures. This Report extended the argument that aerosols might be used to rescue the models, since their effects could cause cooling, and because these effects are so little known they could be used to "adjust" model deficiencies.

The Report also launched a new set of *scenarios* to replace the rather crude number of four scenarios promoted in the First Report, which included the notorious *Business as Usual scenario*. The *Business as Usual* scenario never really died, because its extreme assumptions are a favourite of Government economists and failed U.S. Presidential candidates.

The new scenarios, labelled IS92a.b.c.d.e.f were described in more detail in a supplementary Report.[92]

I published a critical appraisal of both the 1990 and 1992 Reports[93] and another *The IPCC Scenarios are they Plausible*[94] which suggested that most of them were not. Castles and Henderon also criticised them for their statistics.[95] The public and the politicians always chose the least plausible and the most pessimistic for their comments. These scenarios were used for the subsequent IPCC Report until they were replaced by the scenarios described in their 2001 Report.[96]

The Chapter on *Scenarios* states:

> *Scenarios are not predictions of the future and*

---

[92] Pepper, W., Leggett, J., Swart, R., Wasson, J., Edmonds, J., Mintzer, I., May 1992, *Emissions Scenarios for the IPCC: An Update*, Published by the IPCC Cambridge University Press, page 104

[93] Gray, V. R., 1992, *The IPCC Report on Climate Change* and *The 1992 IPCC Supplement: An Appraisal*, in *The Greenhouse Debate Continues*, edited S. Fred Singer, Science and Environment Policy Project; ICS Press San Francisco

[94] Gray, V. R., 1998, *The IPCC Scenarios; are they Plausible?*, 1998, Climate Research 10, 155-162

[95] Castles, I. and Henderson, D., 2003, *The IPCC Emissions Scenarios: An Economic Statistical Critique,* Energy and Environment 14, (2&3) 159-185

[96] Nakicenovic, N. and Swart, R., editors, 2000, *IPCC Special Report: Emissions Scenarios*, Cambridge University Press

should not be used as such.

Repeated statements such as this one by the originators, that scenarios should not be used for forecasts have been routinely ignored by politicians, the media and Governments without a single protest from any IPCC official.

This Report was the first one to which I made comments on both of the circulated drafts. At the time, drafts were sent to interested parties by the New Zealand Ministry of the Environment and the comments were collected by them to forward to the IPCC. My comments were sent by Rob Whitney, from the NZ Coal Research Association.

The comments were regarded as secret and we were never told what happened to them. I had started a series of Greenhouse Bulletins which were originally briefings to the Coal Research Association, but soon became both local and international with the development of email. I devoted two issues of my Bulletin trying to find out whether they had taken any notice of our comments. The results were disappointing.

Climate Change 1994

Climate Change 1994 from the IPCC was a combination of two topics: *Radiative Forcing of Climate Change* and *An Evaluation of the IPCC IS92 Emission Scenarios.* It was provided to support the coming into force of the *Framework Convention on Climate Change* on the 21st March 1994.

The first part was from the first IPCC Committee WGI (Science) and the second part was from the third IPCC Committee (Impact), WGIII.

The Contents are as follows:

Foreword
Part 1

The Global Warming Scam

Preface to WGI Report
Dedication (to Ulrich Siegenthaler)
Summary for Policymakers: Radiative Forcing of Climate Change.
1. CO_2 and the Carbon Cycle
2. Other Trace Gases and Atmospheric Chemistry
3 Aerosols
4 Radiative Forcing
5 Trace gas Radiative Forcing Indices
Part II
Preface to WGIII Report
Summary for Policymakers: An Evaluation f the IPCC 1992 Emission Scenarios
6. An Evaluation of the IPCC Emission Scenarios
Appendix 1 Organisation of the IPCC
Appendix 2 List of Major IPCC Reports
Appendix 3 Contributors to the WGI Report
Appendix 4 Reviewers of the WGI Report (I am named for the first time under "Non-Governmental Organisations)
Appendix 6 Acronyms
Appendix 7 Units
INDEX

There are now even more *Authors* plus *Contributors*. Chapters 1 and 2 also have *Modellers*.

Part 1 introduces the topic of *Global Warming Potential* which enables them to treat all greenhouse gases (except, of course, *water vapour*) as if they behave like carbon dioxide.

In Part II there is the statement:
Since scenarios deal with the future they cannot

be compared with observations...

This absurd statement fails to recognize that the future has an unwelcome tendency to happen and when it does, the scenarios can indeed be checked and as has happened, found wanting.

So they do not need to check whether any of them actually happen, and they tend to prefer *projections* so far ahead nobody can check.

Climate Change 1995: The Science of Climate Change

Climate Change 1995 was the second major Report of the IPCC. It was prepared to launch the first meeting of the *Conference of the Parties (COP 1)* of the signatories of the *Framework Convention on Climate Change,* in Berlin from 20th March to April 7th 1995. It was also used for the subsequent meetings of COP 2, 8-10 July in Geneva, and COP 3 December 1st to 10th in Kyoto, where the *Kyoto Protocol* which imposes compulsory restrictions of greenhouse gas emissions on all signatories of the FCCC, was launched.

The Chapters were as follows:

>*Foreword*
>*Preface*
>*Summary for Policymakers*
>*Technical Summary*
>1. *The Climate System: An Overview*
>2. *Radiative Forcing of Climate Change*
>3. *Observed Climate Variability and Change*
>4. *Climate Processes*
>5. *Climate Models: Evaluation*
>6. *Climate Models – Projections of Future*

Climate
   7. *Changes in Sea Level*
   8. *Detection of Climate Change and Attribution of Causes*
   9. *Terrestrial Biotic Responses to Environmental Change and Feedbacks to Climate*
   10. *Marine Biota Responses to Environmental Change and Feedbacks to Climate*
   11. *Advancing our Understanding*
   Appendix 1. *Organisation of the IPCC*
   Appendix 2. *List of Major IPCC Reports*
   Appendix 3. *Contributors to Climate Change 1995: The Science of Climate Change* (530, including duplicates)
   Appendix 4. *Reviewers* (557 including duplicates. My name is included under "Non-Governmental Organisations," and spelled wrong).

Chapter1 has only three authors: Trenberth, Houghton and Meira Filho. The rest have large numbers of authors and contributors.

The *Summary for Policymakers* is stated to have been approved in detail at the Madrid meeting 27-29 November 1995.

There is now a *Technical Summary* as well as a *Summary for Policymakers* to save people the chore of actually reading the Report. The authors of both of these are not revealed, but it is claimed that the *Technical Summary* is not approved in detail.

As before, there is no index.

The *Observations* have been moved up to number 3, and they no longer place the performance of their models at the beginning. However, Chapter 1 (*The Climate System*) and Chapter 2 (*Radiative Forcing*) are still prominent.

I could claim a major improvement. The First Draft of the 1995 Report had a Chapter 5 *Validation of Climate Models* as in the First Report. I pointed out that it was wrong since no climate model has ever been validated and they did not even try to do so. They thereupon changed the word *Validation* to *Evaluation* no less than fifty times in the Second Draft and have used it exclusively ever since.

Validation is a term used by computer engineers to describe the rigorous testing process that is necessary before a computer-based model can be put to use. It must include successful prediction over the entire range of circumstances for which it is required. Without this process it is impossible to find out whether the model is suitable for use or what levels of accuracy can be expected from it. Without comprehensive validation a model us useless and should be ignored and discarded.

The IPCC has never attempted this process and they do not even discuss ways in which it may be carried out. As a result, the models are worthless. Their possible inaccuracy is completely unknown. The IPCC has developed an elaborate procedure for covering up this deficiency which is well described in the IPCC document on *Guidance Notes for Lead Authors on Addressing "Uncertainties"*. It includes attempts to *simulate* those past climate sequences where suitable adjustment of the uncertain parameters and equations in their models can be made to give an approximate fit, but they rely largely on the elaborate procedure for mobilizing the opinions of those who originate the models. Most of them depend financially on acceptance of the models, so their opinions are handicapped by their conflict of interest.

They never try to grade the different models. Since they are never validated, they have no means of doing it. Also, they dare not antagonise any of the modelists by placing them in any order. They try to get out of this by carrying out intercomparison sessions which can be used to lump the different models into groups. But they would never admit that, perhaps, one group is better than another.

From the 1995 Report on, the IPCC always makes *projections*, never *predictions*. They thus admit that their models are not suitable for prediction at all.

Also, everything is *evaluated*, but not *validated*. There can never be preferred models or scenarios, as they have no way of choosing between them.

Almost all the opinions expressed are based on assuming that a correlation implies a cause and effect relationship. This defies a fundamental logical principle, but it is evaded by calling the process *attribution*.

Since the alternative explanations are always marginalized or distorted, *attribution* to *anthropogenic change* always wins. Very little credence is given to anthropogenic changes that do not involve greenhouse gas emissions, such as land use and urban changes.

The 1995 Report concluded:

> *The balance of the evidence suggests a discernible human influence on global climate.*

This is a typical example of what I have called *Doublespeak*.

Surely all organisms, including humans, have some sort of influence on the climate, and it is usually discernible, so why was it necessary to carry out such an elaborate exercise to find that the balance of their evidence suggests such an obvious fact?

It does not mention *greenhouse gases* at all, but it is obviously intended to justify those who support the theory that the climate is controlled by greenhouse gases, in interpreting it as a support for their views.

Up to this Report it was claimed that they were entirely the work of Working Group I apart from the early statement that they had omitted a *minority opinion*. All the same, there remained reservations about whether the claims made could be due to natural causes.

They claim that this entire Report was written by *Working Group I*, but the Introduction to the *Summary for Policymakers* makes it clear that this was the only part that was *approved by subjecting to line by line discussion and agreement* in a plenary meeting. The rest were only accepted but not *approved in detail*.

It soon became evident that the Government Representatives, who attended only the final plenary session, were not happy with the rest of the Report and intended to show who was boss.

All the Reports suffered from the problem which arises by agreeing the *Summary for Policymakers* after the Final Version of the Main Report has been produced. Since the conclusions of the main Report did not agree with the Government Approved *Summary,* one of the scientists (Ben Santer) had the thankless task of altering statements in the full report to coincide with the *Summary*.[97] The details of the changes to Chapter 8 (*Detection of Climate Change and Attribution of Causes*) are as follows.[98]

The original Working Group I report was approved by the IPCC in December, 1995. Subsequent to that approval, IPCC has carried out additional edits to the document. Some changes are editorial, serving to add clarification or to correct sentence structure. However, other changes appear to go beyond that and have the effect of changing the substance and tone of this chapter. The most significant edits are identified below. New material is italicized while deleted material has a strike through it.

---

[97] *Reclaiming Climate Science*, http://www.greenworldtrust.org.uk/Science/Social/IPCC-Santer.htm

[98] Ibid.

The Global Warming Scam

Summary

"~~Many but not all~~ The Majority of these studies show that the observed changes in global-mean, annually-averaged temperature over the last century is unlikely to be due entirely to natural fluctuations of the climate system."
deleted:

> The evidence rests heavily on the reliability of the (still uncertain) estimates of natural variability noise levels.

new:

> Furthermore, the probability is very low that these correspondences could occur by chance as a result of natural internal variability. The vertical patterns of change are also inconsistent with the response patterns expected for solar and volcanic forcing.
>
> Viewed as a whole, these results indicate that the observed trend in global ~~warming~~ mean temperature over the past 100 years is ~~larger than our current best estimates of natural climate variations over the last 600 years.~~ unlikely to be entirely natural in origin.

Section 8.1

> The attribution of a detected climate change to a particular causal mechanism ~~can be established only by testing~~ involves tests of competing hypotheses.

115

The claimed statistical detection of an anthropogenic signal in the observations must always be accompanied by the caveat that other explanations for the detected climate-change signal cannot be ruled out completely, ~~unless a rigorous attempt has been made to do so.~~

new:

There is, however, an important distinction between achieving 'practically meaningful' and 'statistically unambiguous' attribution. This distinction rests on the fact that scientists and policymakers have different perceptions of risk. While a scientist might require decades in order to reduce the risk of making an erroneous decision on climate change attribution to an acceptably low level (say 1-5%), a policymaker must often make decisions without the benefit of waiting decades for near-statistical certainty.

Section 8

We now have: more relevant model simulations, both for the definition of an anthropogenic climate change signal* ~~and for the estimation of natural internal variability~~.
* *more relevant simulations for the estimation of natural internal variability, and initial estimates from paleoclimatic data of total natural variability on global or hemispheric scales; more powerful statistical methods for detection of anthropogenic change,* ~~and a better understanding of simpler statistical methods~~ *and*

increased application of pattern-based studies with greater relevance for attribution.

Section 8.2.2 Inadequate Representation of Feedbacks

new:

Deficiencies in the treatment and incorporation of feedbacks are a source of signal uncertainty.

Section 8.2.5

Current pattern-based detection work ~~has not attempted~~ is now beginning to account for these forcing uncertainties.

Section 8.3.2

Initial attempts are now being made ~~For these reasons and many others, scientists have been unable to use paleoclimate data in order~~ to reconstruct a satisfactory, spatially-comprehensive picture of climate variability over even the last 1,000 years. ~~Nevertheless~~, The process of quality-controlling paleoclimatic data, integrating information from different proxies, and improving spatial coverage should be encouraged. ~~Without a~~ Better paleoclimatic data bases for at least the past millennium, ~~it will be difficult~~ are essential to rule out natural variability as an explanation for recent observed changes, ~~or~~ and ~~to~~ validate coupled model noise estimates on century time scales (Barnett et al., 1995).

Section 8.3.3.3

deleted:

While such studies help to build confidence in the reliability of the model variability on interannual to decadal time scales, there are still serious concerns about the longer time scale variability, which is more difficult to validate (Barnett et al., 1995). Unless paleoclimatic data can help us to 'constrain' the century time scale natural variability estimates obtained from CGCMs, it will be difficult to make a convincing case for the detection and attribution of an anthropogenic climate change signal.

Section 8.4.1

deleted:

While none of these studies has specifically considered the attribution issue, they often draw some attribution-related conclusions, for which there is little justification.

Section 8.4.1.1

The conclusion that can be drawn from this body of work, and earlier studies reported in Wigley and Barnett (1990) is that the warming trend to date is unlikely to have occurred by chance due to internally-generated variability of the climate system, ~~although this explanation cannot be ruled out. This, however, does not preclude the possibility that a significant part of the trend is due to natural forcing factors.~~ Implicit in such studies is a weak attribution statement--i.e., some (unknown) fraction of the observed trend is being attributed to human influences. Any such attribution-related conclusions, however, rest heavily on the reliability of our estimates of both century time-scale natural variability and the magnitude of the observed global warming mean trend. At best, therefore, trend significance can only provide ~~provides~~ circumstantial support for the existence of an anthropogenic component to climate change, ~~but does not directly address the attribution issue.~~

Section 8.4.1.3

These empirical estimates of ~~In summary, such studies offer support of~~ a DT2x are subject to considerable uncertainty, as shown in a number of studies (see, e.g., Wigley and Barnett, 1990; Wigley and Raper, 1991b; Kheshgi and White; 1993b). In summary, such studies offer support for a DT2x value similar to that obtained by GCMs, and suggest that human activities have had a measurable impact on global climate, but they cannot ~~help~~ to establish a unique link between anthropogenic forcing changes and climate change.

Vincent Gray

Section 8.4.2.1

new:

> *Implicit in these global mean results is a weak attribution statement—if the observed global mean changes over the last 20 to 50 years cannot be fully explained by natural climate variability, some (unknown) fraction of the changes must be due to human influences.*

deleted:

> *None of the studies cited above has shown clear evidence that we can attribute the observed changes to the specific cause of increases in greenhouse gases.*

Section 8.4.2.3.

new:

> *To date, pattern-based studies have not been able to quantify the magnitude of a greenhouse gas or aerosol effect on climate. Our current inability to estimate reliably the fraction of the observed temperature changes that are due to human effects does not mean that this fraction is negligible. The very fact that pattern-based studies have been able to discern sub-global-scale features of a combined CO_2 + aerosol signal relative to the ambient noise of natural internal variability implies that there may be a non-*

negligible human effect on global climate.

Section 8.5.2

new:

Simultaneous model-observed agreement in terms of changes in both global means and patterns, as in the recent study by Mitchell et al. (1995a), is even less likely to be a chance occurrence or the result of compensating model errors.

Section 8.6

Finally we come to the ~~most~~ difficult question of ~~all: 'When will the detection and unambiguous attribution of human-induced climate change occur?'~~ when the detection and attribution of human-induced climate change is likely to occur. The answer to this question must be subjective, particularly in the light of the very large signal and noise uncertainties discussed in this Chapter, ~~it is not surprising that the best answer to this question is 'We do not know'.~~ Some scientists maintain that these uncertainties currently preclude any answer to the question posed above. Other scientists would and have claimed, on the basis of the statistical results presented in Section 8.4, that confident detection of a significant anthropogenic climate change has already occurred. ~~would and have claimed, on the basis of the results presented in Section 8.4, that detection of a significant climate change has~~

~~already occurred.~~ As noted in Section 8.1, attribution involves statistical testing of alternative explanations for a detected observed change and Few ~~if any~~ would be willing to argue that completely unambiguous attribution of (all or part of) this change ~~to anthropogenic effects~~ has already occurred, or was likely to happen in the next several years.

new:

However, evidence from the patterned-based studies reported on here suggests that an initial step has now been taken in the direction of attribution, since correspondences between observations and model predictions in response to combined changes in greenhouse gases and anthropogenic sulphate aerosols:

have now been seen both at the surface and in the vertical structure of the atmosphere;

have been found in terms of complex spatial patterns rather than changes in the global mean alone;

show an overall increase over the last 20 to 50 years;

are significantly different from our best model-based estimates of the correspondence expected due to natural internal climatic variability.

Furthermore, although quantitative attribution studies have not explicitly considered solar and volcanic effects, our best information indicates that the observed patterns of vertical temperature change are not consistent with the responses expected for these forcings.

The body of statistical evidence in Chapter 8, when examined in the context of our physical understanding of the climate system, now points toward a discernible human influence on global climate. Our ability to quantify the magnitude of this effect is currently limited by uncertainties in key factors, including the magnitude and pattern of longer-term natural variability and the time-evolving patterns of forcing by (and response to) greenhouse gases and aerosols.

Section 8.7

APPARENTLY DELETED!

This problem has been reduced in subsequent Reports by the use of elaborate "guidelines" which the Lead Authors are expected to impose on all contributors. It is reproduced as an Appendix to the Report.

The 1995 Report let in some disagreement in the Chapter entitled "Climate Processes," which included R. S. Lindzen, who is a prominent critic of the whole process, and it did develop the general theme that the models were far more inaccurate than is generally assumed. This happened also in the 2001 Report, but it has been eliminated from the 2007 Report.

The Special Report on Emissions Scenarios 2000

This Report was produced by a sub-committee of WGIII. The Drafts of this Report were circulated only to economists and environmental activists. I can claim to have been the only scientist to have commented on the second draft, as its existence came to my notice and I was permitted to borrow the copy from the New Zealand Ministry of Environment. I had a deadline of only one week, but I made copious comments, most of which were, of course, rejected.

The *projections* of the IPCC are a combination of computer climate models (which have never been validated) and scenarios of what might happen in the future. There have up to now been three sets of these, the SA series from the First Report, the IS90 series from the 1992 Supplement Report, and now the SRES series which were launched by the 2000 Report which was prepared by a sub-committee of the WGIII (Impacts) committee of the IPCC. This committee was staffed mainly by environmental enthusiasts committed to exaggerate future change. Their Report was not circulated to scientists for comment or to experienced professional economists, so its exaggerated scenarios were imposed on the scientists of the 2001 and 2007 Reports in order to boost the projections of those reports.

I can give a personal experience of how this happened. The First Draft of the 2001 Report had a maximum projected global temperature rise for the year 2100 of 4°C. The next draft raised this to 5.8°C by inventing a new scenario (A1F1) and using many models, including a drastic one. The politicians must simply have issued a demand to do so.

The scenarios have been criticized by Castles and Henderson.[99]

---

[99] Castles, I. and Henderson, D., 2003, *The IPCC Emissions Scenarios: An Economic Statistical Critique,* Energy and Environment 14, (2&3) 159-185.

Climate Change 2001: The Scientific Basis

This 2001 Report[100] is the one I discussed in some detail in my book *The Greenhouse Delusion: A Critique of Climate Change 2001*[101] It claims to be:

> ***Based on a draft prepared** by a long list of **contributors** plus many **authors** and **reviewers**. But it still fails to mention that it had to be approved line by line by the anonymous Government Representatives who control the IPCC, After the scandal of the previous report strong measures were taken to ensure that Lead Author and review comments had to be approved by the Government representatives.*
>
> *The Chapters were as follows:*
> *Foreword*
> *Preface*
> *Summary for Policymakers*
> *Technical Summary*
> *1. The climate System: An Overview*
> *2. Observed Climate Change and Variability*
> *3. The Carbon Cycle and Atmospheric*

---

[100] Houghton, J. T., Ding, Y., Griggs, D.J., Noguer, M., van der Linden, D.J., Dai, X., Maskell, K., and Johnson, C. A , editors, 2001.*Climate Change 2001: The Scientific Basis*, Cambridge University Press

[101] Gray, V. R., 2002, *The Greenhouse Delusion: A Critique of Climate Change 2001,* Multi-Science publishers

Carbon Dioxide

4. Atmospheric Chemistry and Greenhouse gases

5. Aerosols, their Direct and Indirect Effects

6. Radiative Forcing of Climate Change

7. Physical Climate Processes and Feedbacks

8. Model Evaluation

9. Projections of Future Climate Change

10. Regional Climate Information- Evaluation and Projections

11. Changes in Sea Level

12. Detection of Climate Change and Attribution of Causes

13. Climate Scenario Development

14. Advancing our Understanding

Appendix I Glossary

Appendix II SRES Tables

Appendix III Contributors to the Report (15 pages, approximately 750)

Appendix IV Reviewers of the Report (11 pages. Approximately 550. I get included under "New Zealand")

Appendix V Acronyms and Abbreviations

Appendix VI Units

Appendix VII Some Chemical Symbols used in this Report

Appendix VIII Index

The "Summary for Policymakers" is "Based on a draft prepared by" over 50 authors.

The "Technical Summary" has defined authors, but it is "accepted" but not "approved" by Working Group I.

This time, there is an Index.

Apart from Chapter 1 the Chapters now have Coordinating

The Global Warming Scam

Lead Authors, Lead Authors, Contributors and Review Editors.

The *Summary for Policymakers* is based on a draft prepared by a very large number of authors, but after the scandal of the last Report it would have been approved line by line by the Government Representatives.

The *Observations* Chapter has moved up to #2 and *Radiative Forcing* moved down to #6, but the rest are otherwise unchanged.

Chapter 1, *The Climate System, an Overview* had a Coordinating Lead Author, three Lead Authors and two Review Editors, one of whom was B Bolin.

On page 97, it stated:

> *The fact that the global mean temperature has increased since the late 19th century and that other trends have been observed does not necessarily mean that an anthropogenic effect on the climate has been identified. Climate has always varied on all time-scales, so the observed change may be natural.*

However, this was followed by a more detailed analysis is required to provide evidence of a human impact.

In the *Summary for Policymakers* they seem to think that this evidence had now been supplied.

> *There is new and stronger evidence that most of the warming observed over the last 50 years is attributable to human activities*

and...

> *...in the light of the new evidence and taking*

> *into account the remaining uncertainties, most of the observed warming over the last 50 years is likely to have been due to the increase in greenhouse gas concentrations*

By selecting *the last 50 years*, this statement deliberately ignored the temperature records from weather balloons (for the last 41 years) and satellites (for the past 21 years) which did not show evidence of significant warming. They were therefore selecting the combined weather station and ship record, with all its obvious bias, as their only evidence of *observed warming*.

Since that record shows an *observed cooling* between 1950 and 1975, half of the selected range, the statement makes little sense. What they are really saying is that the *observed warming* in the combined surface record has only been evident for the past 25 years, without mentioning that it disagrees sharply with the observed temperature record from weather balloons and satellites over the same period. They seem to assume that the *greenhouse effect* only operates for 50 years.

Climate Change 2007: The Physical Science Basis

The fourth major IPCC Report was prepared for the meeting of COP 13 at Nusa Dua, Bali from 3rd-14th December 2007.[102]

The following are the Chapters of the Fourth IPCC Major WGI Report:

> *Foreword*
> *Preface*
> *Summary for Policymakers*

---

[102] Solomon, S., Qin, D., Manning, M.R., Marquis, M., Averyt, K., Tignor, M.H., Miller, H.L., and Chin, Z., editors, *Climate Change 2007: The Physical Science Basis (IPCC)*, Cambridge University Press

The Global Warming Scam

Technical Summary
1.Historical Overview of Climate Change Science
2.Changes in Atmospheric Constituents and Radiative Forcing
3.Observations Atmosphere, Surface and Climate Change
4.Observations Changes in Snow Ice and Frozen Ground
5 Observations Ocean Climate Change and Sea Level
6 Paleoclimate Coupling Between Changes in The Climate System and Biogeochemistry
7.Climate Models and Their Evaluation
8.Understanding and Attributing Climate Change
9.Global Climate Projections
10.Regional Climate Projection
Annex 1 Glossary
Annex II Contributors
Annex III Reviewers
Annex IV Acronyms
Index

The *Summary for Policymakers* now has a list of *Drafting Authors*, making it plain that they are taking dictation from the un-named government representatives.

The Technical Summary is once more Accepted, but not approved in details. Its authors are the same as the Drafting Authors of the Summary for Policymakers.

Because of the ambiguous statement of Chapter 1 of the last Report there is a completely different Chapter 1 entitled *Historical Overview of Climate Change Science* which is a highly selective history boosting the activities of the IPCC. This Chapter and all of the others have now acquired *Coordinating Lead Authors, Lead Authors, Contributors* and *Review Editors*.

One of its features is to conceal the very existence of measurements of atmospheric carbon dioxide concentration before 1958 which have been documented by Beck. These show many values that are higher than those claimed today and a variability which would interfere with the IPCC calculations of *radiative forcing*. More than 90,000 accurate chemical analyses of CO_2 in air since 1812 published in peer reviewed Journals by recognised scientists have been eliminated from climate history.[103][104]

The Chapters in *Climate Change 2001* are only slightly rearranged from the previous Report. The key claim of *Climate Change 2007* is that:

> *Most of the observed increase in globally averaged temperature since the mid-20th century is **very likely** due to the observed increase in anthropogenic greenhouse gas concentration.*

As with the previous Report, the main *observed* temperature records which disagree with their opinion are those from weather balloons, which begin in 1958 and those from satellites, which begin in 1978. They eliminate them from consideration by selecting the only record showing an increase, the unreliable mean global surface temperature anomaly. Even this record

---

[103] Beck, E-G, 2007, *180 Years of Atmospheric Gas Analysis by Chemical Methods*, Energy and Environment 18 259-281

[104] Beck E-G, *Evidence of Variability of Atmospheric CO_2 Concentration During the 20th Century*,
http://www.biomind.de/realCO2/literature/evidence-var-corrRSCb.pdf

shows only a fluctuation, with a fall from 1950 to 1976, a rise to 1998 and a fall since then. Then all this is merely *very likely*, based on the unsupported opinion of experts with a conflict of interest, as they are paid to say so.

There is enough for enthusiasts to persuade themselves that the *science is settled* plus sufficient qualifications for the IPCC to claim they never said they were certain, when they are eventually proved wrong. Since there had been no *global warming* for the past 8 years, and we were at the time shivering from the cold in New Zealand, and elsewhere, perhaps that day will come soon.

As a response to a request to the British *Freedom of Information Act*, The IPCC published all the comments and names of Reviewers of this Report.[105]

In March 2006, I was invited to the Climate Center in Beijing to give three lectures on my attitude to Climate Change. I had already provided comments to the First Draft of the Report. The Co-Chair of the WGI Committee at the time was Professor Yihui Ding who was the senior scientist at the Beijing Climate Center. I was welcomed by the then Director General Professor Wenjie Dong and by his senior scientist Mme Zong-Ci Zhao who gave me a paper she had published in association with Professor Ding which showed that there was no evidence of a temperature change in China in the past 100 years.

When I was asked to comment on the Second Draft I made a special effort, as I believed I may have a sympathetic hearing from the Co-Chair.

I submitted a total of 1878 comments which was over 16% of the total (11542). But I was disappointed when I subsequently found that Professor Ding was no longer the Co-Chair. He had been replaced by Qin Dehai, the overall Director of the entire

---

[105] Ibid.

Meteorology organisation.

The full list of reviews and replies have been published.[106]

John McLean provided a detailed analysis of this information, but it fails to mention my contribution.

The Summary for Policymakers has been commented on by Gray[107] and by McKitrick et al.[108]

The whole Report has been reconsidered by Idso and Singer.[109]

Climate Change 2013: The Physical Science Basis

As of this writing, 2013 is the latest Report.[110] It contains the following:

> *Dedication:* Bert Bolin (15 May 1925-30 December 2007)

---

[106] IPCC AR4WG1 Comments,
http://ipcc-wg1.ucar.edu/wg1-commentFrameset.html

[107] Ibid.

[108] McKitrick, R, et al. 2007. *Independent Summary for Policymakers; IPCC Fourth Assessment Report*, Fraser Institute, Vancouver, BC, Canada

[109] Idso, C., Singer, S. F., 2009, *Climate Change Reconsidered I and II*, 2009, Science and Environmental Policy Project and Center for the Study of Carbon Dioxide and Global Change
http://heartland.org/sites/default/files/NIPCC%20Final.pdf and
http://heartland.org/media-library/pdfs/CCR-II/CCR-II-Full.pdf

[110] IPCC, 2013: *Climate Change 2013: The Physical Science Basis,* Contribution of Working Group I to the Fifth Assessment Report of the Intergovernmental Panel on Climate Change, Stocker, T.F., D. Qin, G.-K. Plattner, M. Tignor, S.K. Allen, J. Boschung, A. Nauels, Y. Xia, V. Bex and P.M. Midgley, editors, Cambridge University Press, Cambridge, United Kingdom and New York, NY, USA

The Global Warming Scam

The Working Group I contribution to the Fifth Assessment Report of the Intergovernmental Panel on Climate Change (IPCC) 15 May 1925- 30 December 2007)

Climate Change 2013: The Physical Science Basis is dedicated to the memory of Bert Bolin, the first Chair of the IPCC.

Contents
Foreword
Preface
Dedication
Summary for Policymakers
Technical Summary
1 Introduction
2 Observations: Atmosphere and Surface
3 Observations: Ocean
4 Observations: Cryosphere
5 Information from Paleoclimate Archives
6 Carbon and Other Biogeochemical Cycles
7 Clouds and Aerosols
8 Anthropogenic and Natural Radiative Forcing
9 Evaluation of Climate Models
10 Detection and Attribution of Climate Change: from Global to Regional
11 Near-term Climate Change: Projections and Predictability
12 Long-term Climate Change: Projections, Commitments and Irreversibility
13 Sea Level Change
14 Climate Phenomena and their Relevance for Future Regional Climate Change

> *Annex I Atlas of Global and Regional Climate Projections*
> *Annex II Climate System Scenario Tables*
> *Annex III Glossary*
> *Annex IV Acronyms*
> *Annex V Contributors to the IPCC WGI Fifth Assessment Report*
> *Annex VI Expert Reviewers of the IPCC WGI Fifth Assessment Report*
> *Index*

The final draft Report, dated 7 June 2013, of the Working Group I contribution to the IPCC 5th Assessment Report *Climate Change 2013: The Physical Science Basis* was accepted but not approved in detail by the 12th Session of Working Group I and the 36th Session of the IPCC on 26 September 2013 in Stockholm, Sweden.

They even published another final version in February and a Final-Final-Final printed version on June 7th which contains Supplementary material for Chapters 2,4,7,8,10,13,14 and Annex 1.

The Summary for Policymakers now has Drafting Authors and Draft, Contributing Authors, both of course, subject to the line-by-line approval. The rest have Coordinating Lead Authors, Lead Authors, Contributing Authors and Review Editors.

Drafts, reviewer comments and responses are now published.[111] There are 31,422 comments on the Second Draft and this time my modest contribution of 136 was swamped. There are many critical comments, some from unexpected places.

In the Introduction they make this statement:

---

[111] Drafts Reviews and Responses, https://www.ipcc.ch/report/ar5/wg1/

The Global Warming Scam

> *In this Summary for Policymakers, the following summary terms are used to describe the available evidence: limited, medium, or robust; and for the degree of agreement: low, medium, or high. A level of confidence is expressed using five qualifiers: very low, low, medium, high, and very high, and typeset in italics, e.g., medium confidence. For a given evidence and agreement statement, different confidence levels can be assigned, but increasing levels of evidence and degrees of agreement are correlated with increasing confidence.*
>
> *In this Summary for Policymakers, the following terms have been used to indicate the assessed likelihood of an outcome or a result: virtually certain 99-100% probability, very likely 90-100%, likely 66-100%, about as likely as not 33-66%, unlikely 0-33%, very unlikely 0-10%, exceptionally unlikely 0-1%. Additional terms (extremely likely: 95-100%, more likely than not >50-100%, and extremely unlikely 0-5%) may also be used when appropriate. Assessed likelihood is typeset in italics, e.g., very likely.*

This is the elaborate scam they have established to cover up the very great levels of uncertainty which should be applied to the models (see Chapter 7) They have found that if they express their *confidence* that this proposition is true in these elaborate terms it will somehow replace actual evidence.

They have found it to be necessary to raise this level of *confidence* with each report even when it becomes more and more obvious that their models do not work.

The report is infested with claims that almost everything is *Very Likely*, a term which indicates 95% certainty. The last 5% still has to be left, just in case one day they will have to swallow their words.

My comment was as follows:

> *These are merely the opinions of biased "experts". They are not based on scientific studies.*

In the First Paragraph of the SPM is the following confession

> *The Working Group I contribution to the IPCC's Fifth Assessment Report (AR5) considers new evidence of climate change based on many independent scientific analyses from observations of the climate system, paleoclimate archives, theoretical studies of climate processes and simulations using climate models*

My comment was:

> *This paragraph is confused. You seem to have made a lot of 'observations' which show what we all know already, that the climate is "changing", but "evidence" that you can explain it is dependent on "simulations", and "projections" from untested models, neither of which constitute "evidence" while they are incapable of successful future prediction.*

"Observations" are not the same as actual scientific measurements.

They do make measurements, but they conceal them and

package them up into multi-averaged "data" which are slanted to claim support for the cause.

"Simulations" are mere correlations; they do not prove causation.

Their climate models provide only "projections" and not "predictions." The media have failed to notice this distinction.

They say many of the observed changes are unprecedented over decades to millennia.

My comment was:

> *The periods you quote are ridiculously short and many of the observations are dubious.*

The *Coupled Model Intercomparison Project* has reached phase 5. It has provided a standard set of models:

> *I Decadal Hindcasts and Predictions simulations,*
> *II long-term simulations,*
> *III atmosphere-only (prescribed SST) simulations for especially computationally-demanding models.*

In addition to the earlier *Emissions Scenarios*, which they still use, there is a new set based on Representative Concentration Pathways (RCP)[112] which do not prescribe a detailed pathway to an emission reduction goal, but merely provide a pathway to that may be achieved by any method. It should be noted that there is now a pathway which assumes that all greenhouse gas emissions will be reduced to a very low figure.

---

[112] Ibid.

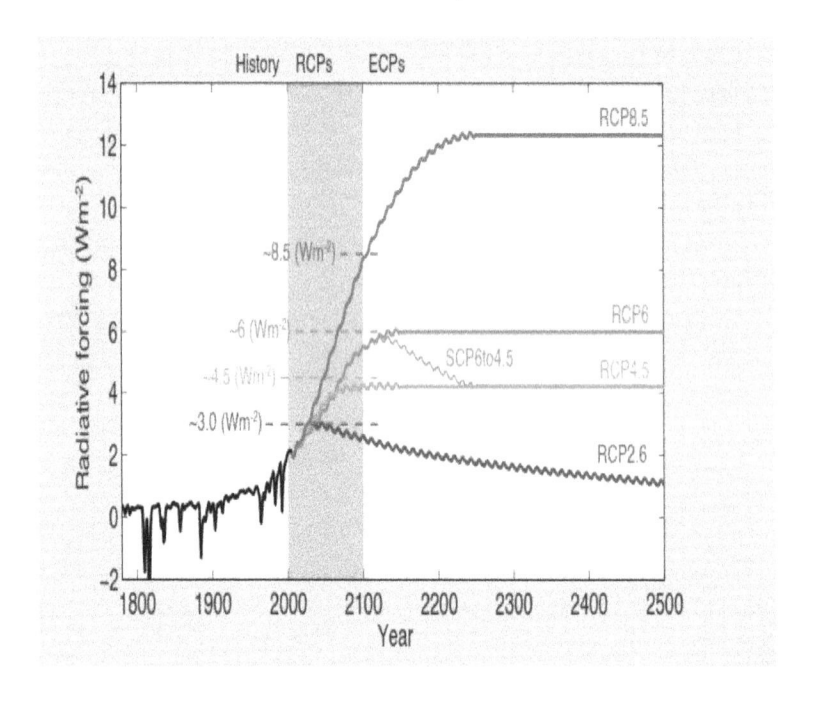

Figure 4.3
Representative Concentration Pathways

Since researchers have used all of the scenarios, they now provide comparisons with all four of them—which is very confusing.

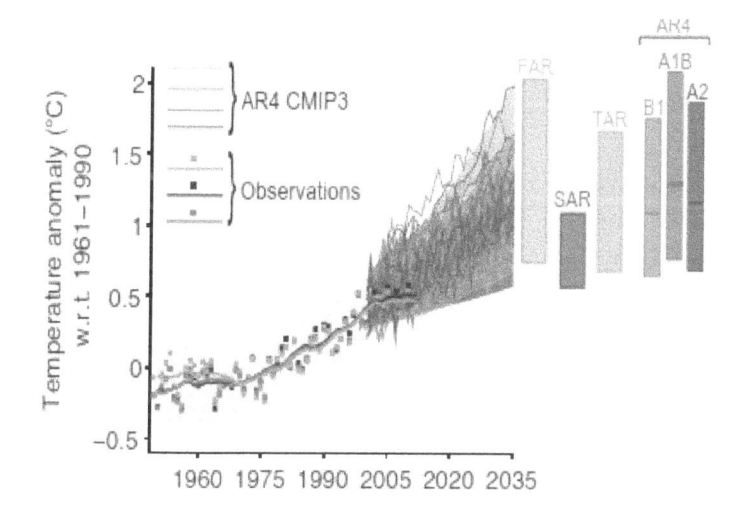

**Figure 4.4
Ranges of IPCC Projections**

The range of projections for temperature in the different IPCC Reports is given in Figure 4.4.

A new feature is Annex 1 which consists of an *Atlas of Global and Regional Climate Projections* covering temperature and precipitation projections compared with the above Representative Concentration Pathways.

This Report has a number of areas where the confidence in the models is low. They include *Increases in intense tropical cyclone activity* and *Increases in intensity and/or duration of drought*.

The main statement of the Summary for Policymakers might be taken as a final conclusion of all the Reports.

> *Warming of the climate system is unequivocal, and since the 1950s, many of the observed changes are unprecedented over decades to millennia. The atmosphere and ocean have*

> warmed, the amounts of snow and ice have diminished, sea level has risen, and the concentrations of greenhouse gases have increased.

What is missing is a claim that a relationship has been established between human emissions of greenhouse gases and any harmful influence on the climate. So why are small levels of warming important, and why does it matter whether greenhouse gas concentrations have increased?

They are desperate to cover up the fact that their surrogate *Annual Mean Global Surface Temperature Anomaly*, which has supplied their main previous argument for warming, has let them down as it has hardly changed for 18 years. Now it is openly claimed to be a genuine global temperature which only increases *decadally* but still with a miserable few decimals of a degree, well below the accuracy of current future forecasts.

The world is scoured to find anything which might have warmed, ignoring the record Antarctic ice growth and persistent cold winters.

Controlling the Scientists

When it was formed the IPCC found itself with large sums of money to recruit scientists to support the climate superscam.

At the time science jobs were scarce and many had to contend with the uncertainties of frequent applications for short term contracts from limited funds.

The IPCC offered a lucrative career with foreign travel, exclusive access to the major journals, academic promotion, and public acclaim, even a Nobel Prize.

They ran into the problem that, scientists are usually trained to think for themselves, and some of those who have been recruited to support the climate change programme find if difficult not to insert their reservations into the opinions that are proscribed for them.

The main mechanism for ensuring uniformity of thought is applied by the presence in all of the IPCC Reports of a *Summary for Policymakers* at the beginning. This is really a Summary BY Policymakers, because it is dictated, line by line by the government representatives who control the IPCC to a group of reliable *Drafting Authors*. It is published before the main Report, to emphasize the need for conformity. In addition they try to exert pressure in the choice of *Lead Authors*, and in the treatment of comments made by the *reviewers* who receive drafts of the Reports.

Despite all this pressure, complete uniformity of thought has, so far, never been achieved.

The First IPCC Report *Climate Change (1990)* stated plainly:

> *The persons named below all contributed to the peer review of the IPCC Working Group I Report. Whilst every attempt was made by the Lead Authors to incorporate their comments, in some cases these formed a minority opinion which could not be reconciled with the larger consensus. Therefore, there may be persons below who still have points of disagreement with areas of the Report.*

But it still stated, even in the *Summary for Policymakers* of the 1990 Report and in its 1992 Supplement:

> *The size of this warming (which they claimed) is broadly consistent with predictions of climate models but it is also of the same magnitude as natural climate variability.*

Climate observations, which appeared only in the last chapter of the 1990 report, were not *"broadly consistent with the predictions of climate models."* Also, all subsequent Reports had to admit that they are actually incapable of making "predictions" but only "projections" dependent on whether you believe the assumptions of the models.

The Second IPCC Report *Climate Change 1995: The Science of Climate Change* had to confront a series of opinions in the Draft of the Final Report which disagreed with the greenhouse theory. It included the following statements:

> *None of the studies cited above has shown clear evidence that we can attribute the observed changes to the specific cause of increases in greenhouse gases.*
>
> *Finally we come to the most difficult question of all: 'When will the detection and unambiguous attribution of human-induced climate change occur?' In the light of the very large signal and noise uncertainties discussed in this Chapter, it is not surprising that the best answer to this question is 'We do not know.'*
>
> *Few if any would be willing to argue that unambiguous attribution of this change to anthropogenic effects has already occurred, or was likely to happen in the next several years.*

One of their scientists (Ben Santer) was given the job of eliminating all the offending passages, or changing them to give a more favoured opinion.

But after all that, they ended up with this equivocal conclusion:

> *The balance of the evidence suggests a discernible*

human influence on the climate.

This is something everybody can agree on. Humans spend all of their efforts in trying to influence the climate. The statement says nothing about greenhouse gases or of carbon dioxide. It supports the offending passages that were deleted.

The Third Report *Climate Change 2001: The Scientific Basis* was the one for which I did a detailed analysis called *The Greenhouse Delusion*.

The following statement appeared in Chapter 1:

> *The fact that the global mean temperature has increased since the late 19th century and that other trends have been observed does not necessarily mean that an anthropogenic effect on the climate has been identified. Climate has always varied on all time-scales, so the observed change may be natural.*

In the Policymakers Summary we get another equivocal opinion:

> *...in the light of the new evidence and taking into account the remaining uncertainties, most of the observed warming over the last 50 years is likely to have been due to the increase in greenhouse gas concentrations.*

Here they change tack. Once again they do not claim that there is evidence this is so, merely that it is the opinion of their paid experts. They seem to think that if they assign the opinion "likely" as meaning greater than 60% probability that this makes it any other than merely an opinion.

And so we come to IPCC Science Report No 4, *Climate*

Change 2007: The Physical Science Basis, where we get a slight amendment to the previous equivocal statement:

> *Most of the observed increase in globally averaged temperature since the mid-20th century is very likely due to the observed increase in anthropogenic greenhouse gas concentrations.*

Again it is not evidence but opinions of their paid experts who are now 95% certain they are right, but it only applies to *most* of the evidence and it only applies to their highly inaccurate temperature series, but not to the more accurate satellite and radiosonde series which began in 1978 and 1958 respectively.

Despite all this they were awarded the Nobel Peace Prize together with Al Gore.

In the Fifth IPCC WGI AR5 Report, *Climate Change 2013: The Physical Science Basis*, we get yet another set of equivocal opinions:

> *Human influence on the climate system is clear.*
> Is this an advance on "discernible"?
> *It is extremely likely that human influence has been the dominant cause of the observed warming since the mid 20th century.*

Again no mention of greenhouse gases—and how much is dominant?

Since the mid 20th Century is pretty short in geological or even in human lifetime terms and it is again arranged so as to eliminate the more reliable satellite and radiosonde measurements.

They ignore completely the *hiatus* that has taken place. Their technique of observing this unequivocal and unprecedented warming has failed to do so for the past 18 years.

The Global Warming Scam

So we have once more the same game they played on the Second Report, except this time it is applied officially by all of them and not by one individual.

They have laboured hard to deal with the *hiatus* by such devices as using *decadal* averages and different starting and ending dates, but not very successfully.

They still avoid the mismatch between *emissions* which are only from the land and *concentrations* which are mainly measured over the sea, and the fact that there is no established relationship between them.

They have deliberately confused *sea level* of the ocean, usually measured by satellites and related to a supposed geoid, calculated from models, and *relative sea level* between sea and land, which is the only one that matters.

It is now obvious that the uncertainties which they have attached to estimates of the earth's energy on their revised diagram are so much larger than the claimed projections of the model calculations that these projections are all are meaningless.

As a final conclusion, attempts of the IPCC to impose rigid discipline on a large group of scientists to persuade them to claim that human emissions of so-called greenhouse gases harm the climate, without being able to supply convincing evidence, has been a failure.

Even their opinions on the supposed reliability of their projections always leave an increasingly small escape route for the day when their approach is proved wrong.

CHAPTER 5: THE GREENHOUSE EFFECT

Jean Baptiste Joseph Fourier

THE GREENHOUSE EFFECT is claimed[113][114] to have been suggested in 1824 by Jean Baptiste Joseph Fourier (1768-1830).[115]

---

[113] Weart, S., 1997, *The Discovery of the Risk of Global Warming*, Physics Today, January, 34-43
[114] Weart, S., 2014, *The Carbon Dioxide Greenhouse Effect*, https://www.aip.org/history/climate/co2.htm#S1
http://www.aip.org/history/climate/co2.htm

In France in 1801, he did experiments on the propagation of heat. In his 1822 book *Théorie Analytique de Chaleur*,[116] he postulated that heat transfer in solids varied according to a constant which became known as the thermal conductivity. He described convection in the following terms:

> *When the heated body is placed in air which is maintained at a sensibly constant temperature, the heat communicated to the air makes the layer of the fluid nearest to the surface of the body lighter; this layer rises more quickly the more intensely it is heated, and is replaced by another mass of cool air. A current is thus established in the air whose direction is vertical, and whose velocity is greater as the temperature of the body is higher.*[117][118]

He does not mention heat transfer by evaporation of water and the release of latent heat in a cooler region.

Fourier attempted to calculate the temperature of the earth in two publications.[119][120] Casey[121] published the English

---

[115] Joseph Fourier, 2014, http://en.wikipedia.org/wiki/Joseph_Fourier

[116] Freeman, A., 1952, *Analytical Theory of Heat*, Translation of Fourier, J. B., 1822, Théorie Analytique de Chaleur in *Great Books of the Western World*, Vol. 45 Encyclopedia Britannica

[117] Ibid. Fourier, *Analytical Theory of Heat*, p. 30

[118] Ibid, Fourier, *Analytical Theory of Heat*

[119] Fourier, J. B. J., 1824, *Remarques Générales Sur Les Températures Du Globe Terrestre Et Des Espaces Planétaires*, Annales de Chimie et de Physique, Vol. 27, pp. 136–167

translation of the 1824 paper by Burgess[122] and an edited version of the paper based on it.[123] He has also provided an edited English translation of the 1827 paper[124] and useful discussion of the errors and misconceptions which have arisen, from which much of the following discussion has been derived.[125]

Fourier was led to his theory from the experiments of his friend de Saussure with his solar heated *hot box*, which was a miniature greenhouse.[126]

Horace-Bénédict de Saussure

Horace-Bénédict de Saussure[127] (1740-1799) was a Swiss physicist who built a solar oven. The increased use of glass

---

[120] Fourier, J. B. J., 1827, *Memoire Sur les Temperatures du Globe Terrestre et des Espaces Planetaires*, Memoires de l'Académie Royale des Sciences, Vol. 7, pp. 569-604

[121] Casey, T., 2014, Text of E. Burgess' 1837 Translation of Fourier, 1824, http://burgess1837.geologist-1011.mobi/

[122] Burgess, E., 1837, *General Remarks on the Temperature of the Terrestrial Globe and the Planetary Spaces*; by Baron Fourier, American Journal of Science, Vol 32, pp. 1-20. Translation from the French, of Fourier, J. B. J., 1824, *Remarques Générales Sur Les Températures Du Globe Terrestre et des Espaces Planétaires*, Annales de Chimie et de Physique, Vol. 27, pp. 136–167

[123] Casey, T., 2014, Fourier (1824) Repaginated with Corrections from Burgess, 1837, http://fourier1824.geologist-1011.mobi/

[124] Casey, T., 2014, English Translation of Fourier, 1827, http://fourier1827.geologist-1011.mobi/

[125] Casey, T., 2014, *The Most Misquoted and Most Misunderstood Science Papers in the Public Domain*, http://geologist-1011.mobi/

[126] *Horace De Saussure and His Hot Boxes*, 2014, http://www.solarcooking.org/saussure.htm

[127] *Horace de Saussure*, 2014, http://en.wikipedia.org/wiki/Horace-B%C3%A9n%C3%A9dict_de_Saussure

during the eighteenth century made many people aware of the hot box's ability to trap solar heat.

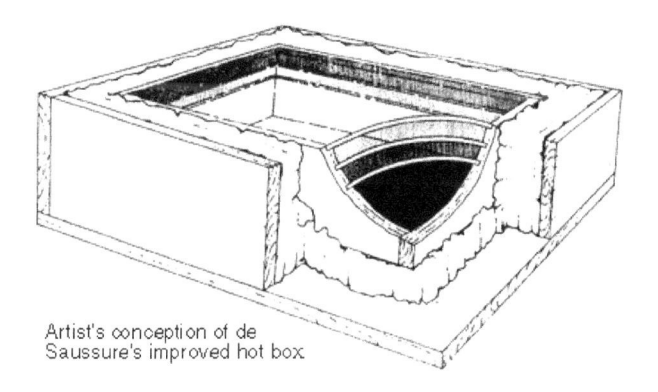

Artist's conception of de Saussure's improved hot box

De Saussure gave a roundabout opinion of how the sun heats a hotbox as follows:

> *Physicists are not unanimous as to the nature of sunlight. Some regard it as the same element as fire, but in the state of its greatest purity. Others envisage it as an entity with a nature completely different from fire, and which, incapable of itself heating, has only the power to give an igneous fluid the movement which produces heat.*[128]

This statement shows how far away from the concepts of modern science was the current understanding of the atmosphere and climate at the time.

Fourier in his 1824 paper (Ibid, page 154) explained the hot box as follows:

---

[128] *Horace de Saussure and his Hot Boxes of the 1700's*, http://solarcooking.org/saussure.htm

The theory of the instrument is easily understood. It is sufficient to remark, 1st, that the acquired heat is concentrated, because it is not dissipated immediately by renewing the air; 2nd, that the heat of the sun, has properties different from those of heat without light. The rays of that body are transmitted in considerable quantity through the glass plates into all the intervals, even to the bottom of the vessel. They heat the air and the partitions which contain it. Their heat thus communicated ceases to be luminous, and preserves only the properties of non-luminous radiating heat. In this state it cannot pass through the plates of glass covering the vessel. It is accumulated more and more in the interval which is surrounded by substances of small conducting power, and the temperature rises till the heat flowing in, shall exactly equal that which is dissipated.

This statement amounts to the following propositions:

- The heated air cannot get out (1st proposition);
- Most of the sun's rays (a considerable quantity) is transmitted through the glass;
- They go to all parts of the vessel;
- The luminous quality of the sun's rays become non luminous heat (infrared);
- Non luminous heat cannot pass through plates of glass;
- It accumulates until the system is in equilibrium.

He would not have been aware of the discovery in 1850 by Melloni that glass absorbs most low temperature infrared radiation.[129]

Fourier applied this behaviour to the atmosphere as follows: (Ibid, page 165).

> *The solar heat has accumulated in the interior of the globe, the state of which has become unchangeable. That which penetrates in the equatorial regions is exactly balanced by that which escapes at the parts around the poles. Thus the earth gives out to celestial space all the heat which it receives from the sun, and adds a part of what is peculiar to itself.*

He thought that there was no net transfer of heat from the sun to the earth, just a bit more in the tropics, that was balanced by loss at the poles. The temperature of the earth was caused by heat received from outer space. This led him to the view that the temperature of space was only just below that of the poles.

He considered that the extra heat needed was supplied from inside the earth which *has accumulated in the interior and is peculiar to itself.*

However, when he added this in, it was not enough, so he had to find some extra heat.

Fourier says:

> *In short, if all the strata of air of which the atmosphere is formed, preserved their density with their transparency, and lost only the*

---

[129] Claude Pouillet, 2011,
http://en.wikipedia.org/wiki/Claude_Pouillet

> mobility which is peculiar to them, this mass of air, thus become solid, on being exposed to the rays of the sun, would produce an effect the same in kind with that we have just described. The heat, coming in the state of light to the solid earth, would lose all at once, and almost entirely, its power of passing through transparent solids: it would accumulate in the lower strata of the atmosphere, which would thus acquire very high temperatures. We should observe at the same time a diminution of the degree of acquired heat...[130]

So, infrared could cause a layer which traps heat just like the glass of de Saussure's hotbox.

Then he says:

> All the terrestrial effects of solar heat are modified by the interposition of the atmosphere and the presence of water. The great motions of these fluids render the distribution more uniform. The transparency of the waters appears to concur with that of the air in augmenting the degree of heat already acquired, because luminous heat flowing in, penetrates, with little difficulty, the interior of the mass, and non-luminous heat has more difficulty in finding a way out in a contrary direction.

So water vapour is responsible for augmenting the degree of heat already acquired.

This view was supported by Claude Pouillet and John

---

[130] Ibid. p. 154

Tyndall, both of whom believed in a warm aether.

Claude Pouillet

Claude Servais Mathias Pouillet (February 16, 1791-June 14, 1868) was a French physicist who developed a *pyrheliometer* and made, between 1837 and 1838, the first quantitative measurements of the Solar constant.[131]

His publication in 1838[132] has been translated into English by Taylor.[133]

Pouillet accepted the view of Fourier that the earth was warmed above the temperature of the aether by the absorption of the sun's rays by the atmosphere. However, by this time, his

---

[131] *Claude Pouillet*, 2011,
http://en.wikipedia.org/wiki/Claude_Pouillet
[132] Pouillet, C., 1838, *Mémoire sur la Chaleur Solaire, sur les Pouvoirs Rayonnants et Absorbants de l'air Atmosphérique et sur la Température de L'espace*, Comptes Rendus des Scéances de l'Academie des Sciences
[133] Pouillet, C., 1838, translated by F. W. Taylor,
http://nsdl.org/archives/onramp/classic_articles/issue1_global_warming/n2-Poulliet_1837corrected.pdf

friend Poisson (the famous statistical mathematician) had calculated that the upper regions of the atmosphere were much cooler than the aether. He carried out experiments at night with an actinometer, an instrument for measuring radiation and from the results calculated that the temperature of space was -142°C.

John Tyndall

John Tyndall (1820-1893) was an Irish-born physicist and mathematician who studied in Germany and from 1853 to 1887 was Director of the Royal Institution in London as the immediate successor of Michael Faraday.[134]

For 12 years from 1859 he carried out a series of studies on the passage of low temperature radiation through a number of gases and vapours.[135]

---

[134] Tyndall, J., 2014, http://en.wikipedia.org/wiki/John_Tyndall
[135] Tyndall, J., 1861, *On the Absorption and Radiation of Heat by Gases and Vapours, and on the Physical Connexion of Radiation, Absorption, Conduction*, The Bakerian Lecture, The London, Edinburgh, and Dublin Philosophical Magazine and Journal of Science, Series 4, Vol. 22, pp. 169-194, 273-285
http://nsdl.org/archives/onramp/classic_articles/issue1_global_warming/n3.Tyndall_1861corrected.pdf.

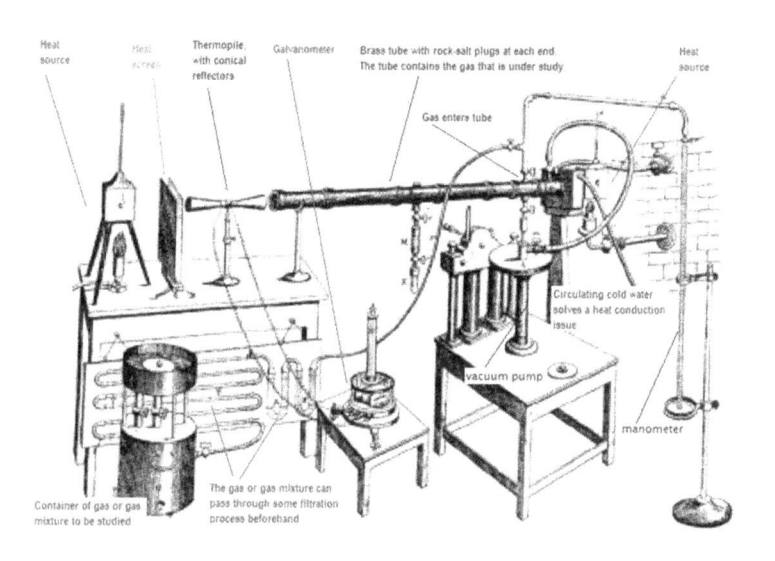

Figure 5.2
Tyndall's Experimental Equipment

Casey has reproduced Tyndall's Bakerian lecture of 1861 which gives details.[136] It is also part of his book *Heat as a Mode of Motion*.[137] [138]

Tyndall was inspired by the recent experiments of Melloni who had studied the infrared behaviour of various gases using a thermomultiplier: a combination of a thermopile and a galvanometer.[139]

---

[136] Ibid.
[137] Tyndall, J., 1868, *Heat as a Mode of Motion*, http://www.archive.org/details/heatamodemotion03tyndgoog
[138] Casey, T., 2014, http://tyndall1861.geologist-1011.mobi/
[139] Melloni, M., 1850, *La Thermochrose, Part 1: Ou La Coloration Calorifique*, http://tinyurl.com/pzyjefa

Tyndall said:

> Melloni found that a glass plate one-tenth of an inch in thickness intercepted all the rays emanating from a source of the temperature of boiling water, and fully 94 percent of rays from a source of 400° Centigrade. Hence a tube closed with glass plates would be scarcely more suitable for the purpose now under consideration, than if its ends were stopped by plates of metal.

De Saussure and Fourier could not have known that glass absorbed most low-temperature radiation.

Tyndall's equipment is illustrated in Figure 5.2.

> A copper cubic container at the right is full of water kept boiling by a flame underneath. The front is coated with lampblack and the radiation passes though a rock salt window and through a brass tube cooled with water. The radiation passes out through another rock salt window and to a double conical device with a thermopile inside it connected to a galvanometer. The tube can be evacuated to give a zero reading for the galvanometer and filled with a gas or vapour to different pressures. Water is circulated around the rock salt to keep it cool. A compensating radiation source is at the far end to enable setting the zero on the galvanometer. He measured the loss of radiation from putting the various gases in the tube.

It has been pointed out by Casey that this arrangement does not measure absorption, a term repeatedly used by Tyndall. It measures relative opacity which is the proportion of radiation passing through the gas. He did not understand that the gas would re-radiate part of the heat as radiation in all directions, some being absorbed by the sides of the tube and some radiating backwards. He seemed to have a rudimentary knowledge of spectroscopy, but his belief in the aether led him to believe in a linear relationship between absorption and concentration. The actual relationship is close to logarithmic.

Tyndall's biographical memoir has the following passage:

> ...he was able to determine the position of aqueous vapour, which, on account of condensation, could not be experimented on directly. Experiments made with dry and humid air corroborated the inference that, as water transcends all other liquids, so aqueous vapour is powerful above all other vapours as a radiator and absorber.
>
> Air sent through the system of drying-tubes and through the caustic-potash tube produced an absorption of about 1.
>
> Air direct from the laboratory, containing therefore its carbonic acid and aqueous vapour, produced an absorption of about 15.
>
> Deducting the effect of the gaseous acids, it was found that the quantity of aqueous vapour diffused through the atmosphere on the day in question, produced an absorption at least equal to thirteen times that of the atmosphere itself.[140]

Tyndall wrote:

---

[140] Tyndall, J., 1903, Biographical Memoir in *Lectures and Essays*. Watts and Company

> It is exceedingly probable that the absorption of the solar rays by the atmosphere, as established by M. Pouillet, is mainly due to the watery vapour contained in the air. The vast difference between the temperature of the sun at midday and in the evening is also probably due in the main to that comparatively shallow stratum of aqueous vapour which lies close to the earth. At noon the depth of it pierced by the sunbeams is very small; in the evening very great in comparison.
>
> The intense heat of the sun's direct rays on high mountains is not, I believe, due to his beams having to penetrate only a small depth of air, but to the comparative absence of aqueous vapour at those great elevations.
>
> But this aqueous vapour, which exercises such a destructive action on the obscure rays, is comparatively transparent to the rays of light. Hence the differential action, as regards the heat coming from the sun to the earth and that radiated from the earth into space, is vastly augmented by the aqueous vapour of the atmosphere.

He believed that *solar rays* are absorbed mainly by water vapour in the atmosphere and it is *far more important than that radiated from earth into space*. So, he is also not responsible for the current concept of the greenhouse effect from carbon dioxide and he did not consider carbon dioxide as important at all.

He also wrote (Ibid, page 277, paragraph 1):

> *De Saussure, Fourier, M. Pouillet, and Mr.*

> Hopkins regard this interception of terrestrial rays as exercising the most important influence on climate.

Tyndall does appear to be the first person to point out that trace gases in the atmosphere are capable of influencing climate. He studied carbon dioxide and methane but he thought water vapour was far more important.

Svante Arrhenius

Svante August Arrhenius (1859-1927) was one of the founders of the science of physical chemistry. He received the Nobel Prize in Chemistry in 1903 for his discovery of ions in aqueous salt solutions.[141] He published several articles on the effects of carbon dioxide on the atmosphere. The one published in English in 1896 was the most influential.[142]

---

[141] *Svante Arrhenius*, 2014, http://en.wikipedia.org/wiki/Svante_Arrhenius
[142] Arrhenius, S., 1896, *On the Influence of Carbonic Acid in the Air upon the Temperature of the Ground*, London, Edinburgh, and Dublin

He made very few measurements himself and the paper depended entirely on his calculations from the measurements by Langley and Very.[143]

Samuel Pierpont Langley

Samuel Pierpont Langley, (August 22, 1834-February 27, 1906) was an American astronomer, physicist and pioneer of aviation.

He is probably best known from the fiasco of his two attempts to launch a man-carrying flying machine across the Potomac River in October and December 1903, both of which failed when the machine plunged directly into the river. He refused to recognise the December 1903 success of the Wright brothers and as Director of the Smithsonian Institution he claimed priority in his museum. The original Wright brothers' flyer was therefore donated to the Science Museum in London where I used to visit as a boy. The Smithsonian claimed Langley's priority when I visited the museum as recently as the 1960s, but

---

Philosophical Magazine and Journal of Science (fifth series), April 1896, vol 41, pages 237-275
[143] Langley, S. P. with Frank W. Very, 1890, *The Temperature of the Moon*, Memoir of the National Academy of Sciences, vol. iv.

then a deal was done and the Wright machine is now in the Smithsonian and the Science Museum has a replica.

Langley invented an instrument, the *bolometer* which could measure the intensity of infrared radiation, which could be used to measure narrow absorption bands of a spectrum supplied by a rock salt prism.

Langley made a series of measurements of the full moon's radiation with this instrument at the Alleghany observatory in 1997.[144] He made measurements at different angles with the moon.

Arrhenius had the idea that by finding out the difference he got from different angles he could calculate the absorption of the moon's radiation by the earth's atmosphere. By assuming that the radiation from the moon was approximately the same as that of the earth he could calculate the absorption of the earth's atmosphere by the trace absorbent gases water vapour and carbon dioxide.

Erren has shown that Langley's measurements used by Arrhenius were preliminary and had serious errors.[145] They became less accurate as they approached the region used by Arrhenius, and they did not reach far enough to include the major absorption frequency of carbon dioxide. He concluded that Langley's observations measured water vapour and not carbon dioxide in the atmosphere.

Arrhenius published many subsequent publications,[146][147][148]

---

[111] Langley, S. P., 2014,
http://en.wikipedia.org/wiki/Samuel_Pierpont_Langley
[145] Erren, H., 2011, *Langley Revisited*,
http://members.casema.nl/errenwijlens/co2/langleyrevdraft2.htm
[146] Arrhenius, S., 1901, *Ueber die Wärmeabsorption durch Kohlensäure*, Annalen der Physik, Vol 4, 1901, pages 690–705

[149] [150] parts of which are available from Erren,[151] [152] who also provides modern information on the moon spectrum,[153] which show that Arrhenius' values were exaggerated[154] [155] and that he did not subsequently amend them.

Erren disagrees with Weart who claimed that the 1901 and 1908 Arrhenius papers lowered his original numbers. He also gives an account of the paper by K. Angstrom which wrongly criticized Arrhenius.[156]

On his first page, Arrhenius states:

---

[147] Arrhenius, S., 1901, *Über Die Wärmeabsorption Durch Kohlensäure Und Ihren Einfluss Auf Die Temperatur Der Erdoberfläche*, Abstract of the proceedings of the Royal Academy of Science, 58, 25–58

[148] Arrhenius, S., 1903, *Lehrbuch der Kosmischen Physik,* Vol I and II, S. Hirschel publishing house, Leipzig

[149] Arrhenius, S., 1906, *Die Vermutliche Ursache der Klimaschwankungen, Meddelanden från K.* Vetenskapsakademiens Nobelinstitut, Vol 1 No 2, pages 1–10.

[150] Arrhenius, S., 1908, *Das Werden der Welten*, Academic Publishing House, Leipzig

[151] Erren, H., 2014, Scanned pages from Arrhenius 1901 http://members.casema.nl/errenwijlens/co2/arrhenius1901/index.html

[152] Erren, H., 2014, *Reply to Weart2* http://members.casema.nl/errenwijlens/co2/arrhweart.htm

[153] Erren, H., 2014, Scanned pages from Arrhenius *Textbook of Cosmic Physics*, 1903, and Arrhenius 1908 *Becoming the Worlds*, http://members.casema.nl/errenwijlens/co2/arrhenius0308/index.html

[154] Erren, H., 2014, *Arrhenius Was Wrong*, http://members.casema.nl/errenwijlens/co2/arrhrev.htm

[155] Erren, H. 2014, Summary Graphs of Arrhenius' Errors, http://members.casema.nl/errenwijlens/co2/arrhenius.html

[156] Erren, H., 2014, Knut Angstrom's Measurements, http://members.casema.nl/errenwijlens/co2/angstrom1900/index.html

> *Fourier maintained that the atmosphere acts like the glass of a hothouse because it lets through the light rays from the sun but retains the dark rays from the ground...*[157]
>
> *...and Langley was by some of his researches led to the view that 'the temperature of the earth under direct sunshine would probably fall to -200°C, if that atmosphere did not possess the quality of selective absorption.*

It should be pointed out that although infrared rays do not pass through the glass of a hot house, they are also not reflected from it, but are absorbed as heat.

Arrhenius says:

> *This view...must be abandoned, as Langley himself in a later memoir, showed that the full moon, which certainly does not possess any sensible heat-absorbing atmosphere, has a 'mean effective temperature' of 45°C.*

Langley's figure for the temperature of the moon was wrong. Today's figure is an average of 107°C. Langley's figures must therefore be treated as completely unreliable, and so are the results calculated from them by Arrhenius.

Arrhenius completely failed to accept that Tyndall had found that water vapour was far more important than carbon dioxide.

He assumed the ratio of carbon dioxide (K) to water vapour (W) in the earth's atmosphere (K/W) using K as 1.5 and W as

---

[157] Ibid.

0.88, a ratio of 1.7.

The concentration of carbon dioxide in the earth's atmosphere is now thought to be 0.039%. The average concentration of water vapour is not known since it varies from place to place from 0 to 4%. If you take 2% as typical, the ratio of water vapour tp carbon dioxide is about 50 to 1.

So, about 98% of Arrhenius' figures and calculations, even if soundly based, still apply to water vapour and not to carbon dioxide.

Arrhenius no longer believed in the aether when he says:

> *Empty space may be regarded as having a temperature of absolute zero.*

He included this basic model of the climate:

> *All authors agree in the view that there prevails an equilibrium in the temperature of the earth and of its atmosphere. The atmosphere must, therefore, radiate as much heat to space as it gains, partly from absorption of the sun's rays, partly through the radiation from the hotter surface of the earth and by means of the ascending currents of air heated by contact with the ground. On the other hand the earth loses just as much heat by radiation into space and to the atmosphere as it gains by absorption of the sun's rays. If we consider a given place in the atmosphere or on the ground, we must also take into consideration the quantities of heat that are carried to this place by means of oceanic or atmospheric currents.*

It may be noticed that Arrhenius regards as important "ascending currents of air heated by contact with the ground" and "the quantities of heat that are carried to this place by means of oceanic or atmospheric currents." In other words: convection and circulation.

He considered the warming effects of increased carbon dioxide as entirely beneficial.

Guy Stewart Callendar

Guy Stewart Callendar[158] (1898-1964) was a steam engineer and inventor who published many studies and articles which revived the claim by Arrhenius that increased atmospheric carbon dioxide heated the earth.

He measured the absorption spectrum of water vapour and carbon dioxide and that of the sky. He ignored water vapour and even believed that radiation was the only form of energy transfer.

---

[158] Callendar, G. S.,
http://en.wikipedia.org/wiki/Guy_Stewart_Callendar

In his 1938 paper he stated:

> If the whole surface of the earth is considered as a unit upon which a certain amount of heat falls every day, it is obvious that the mean temperature will depend upon the rate at which heat can escape by radiation, because no other type of heat exchange is possible.[159]

He seemed to be unaware of conduction convection and evaporation as possible mechanisms of heat transfer.

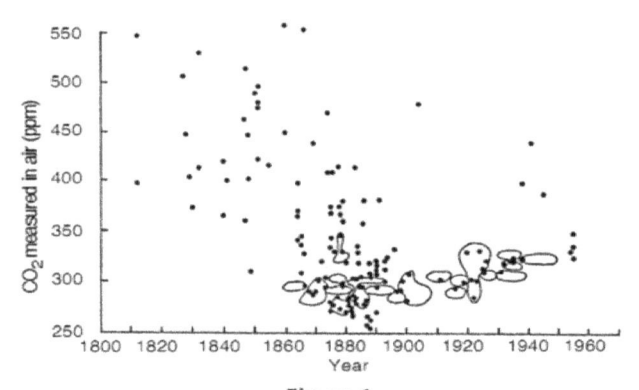

Figure 1
AVERAGE ATMOSPHERIC CO_2 CONCENTRATIONS MEASURED IN THE 19TH AND 20TH CENTURIES

Figure 5.3
Choice of CO_2 Measurements by Callendar

---

[159] Callendar, G. S., 1938, *The Artificial Production of Carbon Dioxide and Its Influence on Climate*, Quarterly Journal of the Royal Meteorology Society, pps. 223-240. http://www.rmets.org/pdf/qjcallender38.pdf

Jarowowski[160] claimed that Callendar chose only those figures for carbon dioxide that suited his theory.

Sir George Simpson

Sir George Simpson, (1878-1965) the eminent meteorologist, who was, at that time, Director of the UK Meteorological Office, in commenting on this paper said:

> *It is not sufficiently realised by non-meteorologists who come for the first time to help the Society in its study that it was impossible to solve the temperature distribution in the atmosphere by working out the radiation. The atmosphere was not in a state of radiative equilibrium, and it also received heat by transfer from one part to another In the second place, one had to remember that the temperature*

---

[160] Jaworowski, Z., 1997, *Ice Core Data show no Carbon Dioxide Increase*, 21st Century Spring, pps 44-52
http://www.21stcenturysciencetech.com/2006_articles/IceCoreSprg97.pdf

distribution in the atmosphere was determined almost entirely by the movement of air up and down. This forced the atmosphere into a temperature distribution which was quite out of balance with the radiation. One could not, therefore, calculate the effect of changing any one factor in the atmosphere, and he felt that the actual numerical results which Mr Callendar had obtained could not be used to give a definite indication of the order of magnitude of the effect.[161]

These remarks continue to be true today.

Thomas Chrowder Chamberlin

Thomas Chrowder Chamberlin (1843-1928) was a respected and influential American geologist and science educator. Chamberlin developed a theory of *climate change* and was one of the first to emphasize carbon dioxide as a major regulator of Earth's temperature, thus anticipating modern *global warming.*

---

[161] Fleming, J. R., 1998, *Historical Perspectives on Climate Change*, Oxford University Press

Chamberlin's graduate seminar at the University of Chicago in 1896 contained all the themes that informed his research programme over the next three decades. These included the carbon dioxide theory of climate change in its relationship to diastrophism and oceanic circulation, the role of water vapour feedbacks in the climate system, and the relationship between multiple glaciations, the climate system, and the formation of the planet.

Hubert Lamb

Hubert Lamb, (1919-1997), distinguished British climatologist; founder of the Climate Research Unit at the University of East Anglia.[162]

He helped launch climatology as an honest personal assessment of properties of local global and ancient climates based on exhaustive scholarship.

Lewin[163] showed that in contrast to current research

---

[162] Lewin, B., *Enthusiasm and Sceptical Science*,
https://enthusiasmscepticismscience.wordpress.com/h-h-lamb/
https://enthusiasmscepticismscience.wordpress.com/2014/02/23/the-scepticism-of-hubert-horace-lamb-part-ii/

directions at CRU, its founding director was an early and vocal climate sceptic.

Against the idea that greenhouse gas emissions were (or would soon be) noticeably warming the planet, Lamb raised objections on many levels. His greatest concern was not so much the lack of science behind the theory; it was how the growing preoccupation with man-made warming was distorting the science.[164]

His successor, an Australian physicist, Tom Wigley, assisted by Phil Jones, have been prominent advocates of the carbon dioxide climate theory.

Gilbert Norman Plass

Gilbert Norman Plass, 1920-2004, was a Canadian physicist who in 1956 published a paper with predictions that the increase in global atmospheric CO_2 levels in the 20th century would affect the average temperature of the earth.[165]

---

[163] Global Warming Policy Foundation, *Hubert Lamb And The Transformation Of Climate Science*
http://us4.campaignarchive1.com/?u=c920274f2a364603849bbb505&id=f3a198d64b&e=199b7f065f

[164] Lamb, H. H., http://www.cru.uea.ac.uk/about-cru/hubert-lamb

[165] Plass, Gilbert N., 1956, *The Carbon Dioxide Theory of Climatic Change*, Tellus 8, 140–154

Roger Revelle

Roger Revelle, (1909-1991), was an American oceanographer who was Director of the Scripps Institution of Oceanography in San Diego from 1950 to 1964. He served as Science Advisor to Interior Secretary during the Kennedy Administration in the early 1960s.

He helped to launch the *International Geophysical Year* (IGY) in 1958 and was founding chairman of the first Committee on Climate Change and the Ocean (CCCO) under the Scientific Committee on Ocean Research (SCOR) and the International Oceanic Commission (IOC). In July 1956, Charles David Keeling joined the SIO staff, and began measurements of atmospheric carbon dioxide at the Mauna Loa Observatory on Mauna Loa, Hawaii, and in Antarctica.

In 1957, Revelle co-authored a paper with Hans Suess[166] that suggested that human gas emissions might cause *global warming* and that bicarbonate chemistry caused a resistance to

---

[166] Revelle, R. & Seuss, H. E., 1987, *Carbon Dioxide Exchange Between Atmosphere and Ocean and the Question of an Increase of Atmospheric CO_2 During the Past Decades*, http://www.odlt.org/dcd/docs/Revelle-Suess1957.pdf

absorption of atmospheric carbon dioxide by the ocean: (the *Revelle Effect*).

Revelle eventually regretted his advocacy of the greenhouse effect.[167]

One of his students, Al Gore, became a propagandist for The Greenhouse Scam with his books and his film *An Inconvenient Truth*.

Robert W. Wood

Robert W. Wood, 1868-1955, was an American physicist and inventor. He wrote a popular textbook called *Physical Optics*.[168]

He presented a theory of the operation of a greenhouse in the Philosophical Magazine in 1909 (Vol. 17, pp. 319-320).[169]

> *XXIV. Note on the Theory of the Greenhouse*
> *By Professor R. W. Wood (Communicated by the*

---

[167] *Uncensored*, http://uncensored.co.nz/tag/global-warming-scam/
[168] Wood, R. W., 1934, *Physical Optics*. Dover publications, New York, 1967
[169] Wood, R. W., 1909, *Note on the Theory of the Greenhouse*, Philosophical magazine, vol 17, p319-320
http://www.wmconnolley.org.uk/sci/wood_rw.1909.html

The Global Warming Scam

Author)

There appears to be a widespread belief that the comparatively high temperature produced within a closed space covered with glass, and exposed to solar radiation, results from a transformation of wave-length, that is, that the heat waves from the sun, which are able to penetrate the glass, fall upon the walls of the enclosure and raise its temperature: the heat energy is re-emitted by the walls in the form of much longer waves, which are unable to penetrate the glass, the greenhouse acting as a radiation trap.

I have always felt some doubt as to whether this action played any very large part in the elevation of temperature. It appeared much more probable that the part played by the glass was the prevention of the escape of the warm air heated by the ground within the enclosure. If we open the doors of a greenhouse on a cold and windy day, the trapping of radiation appears to lose much of its efficacy.

As a matter of fact I am of the opinion that a greenhouse made of a glass transparent to waves of every possible length would show a temperature nearly, if not quite, as high as that observed in a glass house. The transparent screen allows the solar radiation to warm the ground, and the ground in turn warms the air, but only the limited amount within the enclosure. In the "open," the ground is continually brought into contact with cold air by convection currents.

To test the matter I constructed two

enclosures of dead black cardboard, one covered with a glass plate, the other with a plate of rock-salt of equal thickness. The bulb of a thermometer was inserted in each enclosure and the whole packed in cotton, with the exception of the transparent plates which were exposed. When exposed to sunlight the temperature rose gradually to 65°C, the enclosure covered with the salt plate keeping a little ahead of the other, owing to the fact that it transmitted the longer waves from the sun, which were stopped by the glass. In order to eliminate this action the sunlight was first passed through a glass plate.

There was now scarcely a difference of one degree between the temperatures of the two enclosures. The maximum temperature reached was about 55°C. From what we know about the distribution of energy in the spectrum of the radiation emitted by a body at 55°, it is clear that the rock-salt plate is capable of transmitting practically all of it, while the glass plate stops it entirely. This shows us that the loss of temperature of the ground by radiation is very small in comparison to the loss by convection, in other words that we gain very little from the circumstance that the radiation is trapped. Is it therefore necessary to pay attention to trapped radiation in deducing the temperature of a planet as affected by its atmosphere?

The solar rays penetrate the atmosphere, warm the ground which in turn warms the atmosphere by contact and by convection currents. The heat received is thus stored up in the atmosphere, remaining there on account of

> *the very low radiating power of a gas. It seems to me very doubtful if the atmosphere is warmed to any great extent by absorbing the radiation from the ground, even under the most favourable conditions.*
>
> *I do not pretend to have gone very deeply into the matter, and publish this note merely to draw attention to the fact that trapped radiation appears to play but a very small part in the actual cases with which we are familiar.*

Wood showed that internal convection warms the air which cannot escape to be cooled by the outside climate. He does not mention evaporation of water which also cools the surface. In common with Fourier and the others he does not mention what happens at night or when the sun is not present: when the whole frame cools by external convection combined with deposition of dew internally.

The Real Greenhouse

A real greenhouse is a confined sector of the real climate described in Chapter 1. It receives sunlight through glass panels but it restricts atmospheric circulation and shields from precipitation and thus is able to maintain a higher temperature than the outside climate which is cooled by these effects.

Otherwise it behaves in exactly the same way as outside. The sun's rays are absorbed at the base and the frame and raise its temperature. It is cooled when the air above is warmed and rises by convection. It is also cooled when water is evaporated and the air becomes more humid. Eventually all the air in the greenhouse has a higher temperature than the outside, where it would be cooled if escaped. The infrared radiation from the ground and from the atmosphere cannot pass through the glass. But they are not reflected as Fourier and others have surmised, they are absorbed and so heat the surface, then cooled by outside convection and outwards radiation.

As greenhouses are not insulated, the frame is warmed and cooled by the outside air circulation and precipitation. At night or when the sun does not shine, the frame cools by convection and radiation. The air inside also cools but some heat is transferred to the base by deposition of dew when the humidity falls. Internal radiation plays a small but negligible part. Infrared from the base is merely absorbed by the frame and the glass but it is not reflected as Fourier surmised, but the absorbed heat is mainly lost by external convection of the frame.

Summary

The replacement of the accumulated discoveries of meteorology described in Chapter 1 by global climate models based on atmospheric concentrations of carbon dioxide was motivated by an environmentalist delusion that human activity was exclusively responsible for the climate.

The presumed pioneers, Fourier and Pouillet, were only concerned with water vapour. Tyndall showed that water vapour was far more important than carbon dioxide. Yet the wrong greenhouse gas has been chosen, purely because its concentration can be blamed on human activity.

Arrhenius ignored the advice of these pioneers and failed to realise that Langley's measurements did not include carbon dioxide absorption; so his results were for water vapour instead. All subsequent advocates for an important role for carbon dioxide have conveniently failed to realise this.

The Global Warming Scam

CHAPTER 6: THE MODELS SCAM

CLIMATE IS A heat engine with input energy from the sun and exhaust to outer space. It sustains all living organisms by supplying them with necessary energy.

Genuine climate science models are the numerical models used for weather forecasting. They consist of a coupling of the latest mathematical representation of the atmosphere and ocean, into which is substituted experimental observations made with calibrated instruments over definite time periods at every level of the atmosphere. Modern computers permit observations made at ever smaller time and space increments of the climate properties.

Additional essential features of all these models are representations of the position of the sun over each region, plus a range of other local properties and influences such as mountains, events like monsoons, ocean oscillations and the extent of industrialisation. Even after all of this is done, using the most advanced computers as shown in Chapter 1, the accuracy of local of regional temperature forecasts is only ±2°C with a bias of ±1°C and other forecasts are similar.

A global climate may be represented by an assembly of local models where each individual local variability and correction must be preserved. If these are ignored, as is proposed by IPCC climate change models, the accuracy of any forecast is much less than a fully adjusted local model.

An increasing range and variety of climate measurements and observations are now incorporated into weather forecasting models. The importance of water in all its forms is involved in many of them, but so far, there is no use made of measurements of carbon dioxide concentrations or those of other trace gases Indeed, no effort is made to measure concentrations of these gases over most land surfaces.

The IPCC models begin with the assumption that climate is exclusively controlled by the so-called greenhouse gases and that their increase is responsible for widespread harm to the climate, notably warming.

Most scientists would agree that the greenhouse gases have infrared absorption bands which cause a temperature rise in the atmosphere. There is, however, no evidence that this rise is detectable or could be considered as harmful. By contrast, carbon dioxide is responsible for the mere existence of most living organisms and there is evidence any increase is beneficial.

In order to promote their fraudulent climate theory, the IPCC models lump together all of the many observations made by meteorologists into parametric simulations of what they term *natural variability* based on averages of properties over past periods. This task is impossible because no account is made of the chaotic behaviour of all of them and none of the figures possess the necessary symmetry before a scientifically acceptable average or measure of variability can be derived. The diurnal, seasonal and irregular radiant input from the sun imposes heavily skewed populations of parameters. Any attempt of this kind provides the IPCC absurdity of a sun with the same intensity day and night.

Despite these impossibilities, the IPCC models assume that these simulated phoney averages are constants so that the only influence on the climate can be claimed to be provided by increases in human-induced greenhouse gases. Models based on these dubious principles are then combined with scenarios of the future which are dominated by the pipe dreams of committed environmentalists, to provide projections which are assigned levels of likelihood and confidence by scientists whose opinions are closely monitored and controlled by the anonymous government representatives who approve the IPCC Reports.

The models typically calculate the radiative forcing consequences of doubling the effective atmospheric carbon dioxide concentrations (the climate sensitivity) to provide a future projection. This projection must be combined with future scenarios which estimates how long this process will take.

The IPCC has run into trouble with its approved pseudo temperature record which has now been unchanged for 18 years. They have therefore began to project for decades. Their conclusions are not predictions, but projections, which are assigned levels of likelihood by approved and financially dependent IPCC scientists. None of these projections have ever been confirmed.

Their attitude is typically expressed by the following *Frequently Asked Questions*:

> *FAQ 11.1 If You Cannot Predict the Weather Next Month, How Can You Predict Climate for the Coming Decades?*
>
> *Although weather and climate are intertwined, they are in fact different things. Weather is defined as the state of the atmosphere at a given time and place, and can change from hour to hour and day to day. Climate, on the other hand, generally refers to the statistics of weather*

conditions over a decade or more.[170]

They define weather as what climate generally refers to, not what it is.

As discussed in Chapter 1, climate is a general term and weather is the actual condition at a particular place and time. All the same, climate is always essentially local and is different from one place to another. It is perfectly possible over a limited period to provide rough average local weather statistics. Averages with a plausible level of accuracy can only be provided by populations of figures that are approximately symmetrical. This might apply to maximum and minimum daily temperature, wind direction and strength, sunshine hours or rainfall. The frequency of extreme events cannot be judged because of the small numbers and overall climate changes that undoubtedly take place.

---

[170] IPCC, 2013, *Climate Change 2013: The Physical Science Basis. Contribution of Working Group I to the Fifth Assessment Report of the Intergovernmental Panel on Climate Change*, Stocker, T.F., Qin, D., Plattner, G-K, Tignor, M., Allen, S.K., Boschung,J., Nauels, A., Xia, Y., Bex, V., and Midgley, P. M., editors, Cambridge University Press, Cambridge, United Kingdom and New York, NY, USA, 1535 pp., page 964

THE IPCC MODELS

The earliest model is from the first IPCC Report.[171]

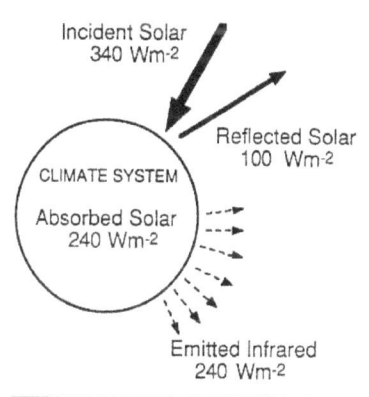

	Absorbed	Emitted
Instantaneous CO2 doubling	240 Wm-2	236 Wm-2
New Equilibrium with no other change	240 Wm-2	240 Wm-2

Figure 6.1
Global Radiation Budget according to Climate Change 1990

Figure 6.1 is extremely crude. It is like the moon. There is no atmosphere, no life, no heat transfer, no rotation and represents an artificial equilibrium.

The maximum solar irradiance is 1370 W/m^2 whereas their

---

[171] Houghton, J. T., Jenkins, G. J. and Ephraums, J. J., editors, 1990 *Climate Change : The IPCC Scientific Assessment*, Cambridge University Press, p. 78

climate system gets only an average of 240 W/m² even on a spherical earth.

The 2nd IPCC Report 1995 even provided a fairly realistic diagram of the climate.[172]

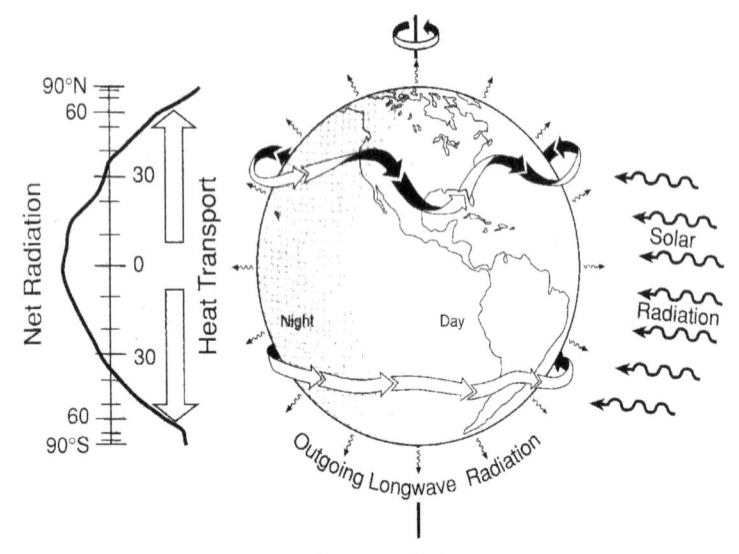

**Figure 6. 2
Diagram of the Climate from the 2nd IPCC Report**

But the very next page provided their real model which differs from Figure 2 in every possible respect.[173]

---

[172] Houghton, J. T., Meira Filho, L. G., Callander, B. A., Harris, N., Kattenberg, A. and Maskell, K., editors, 1996 *Climate Change 1995: The Science of Climate Change*, Cambridge University Press, p. 57
[173] Ibid. p. 58

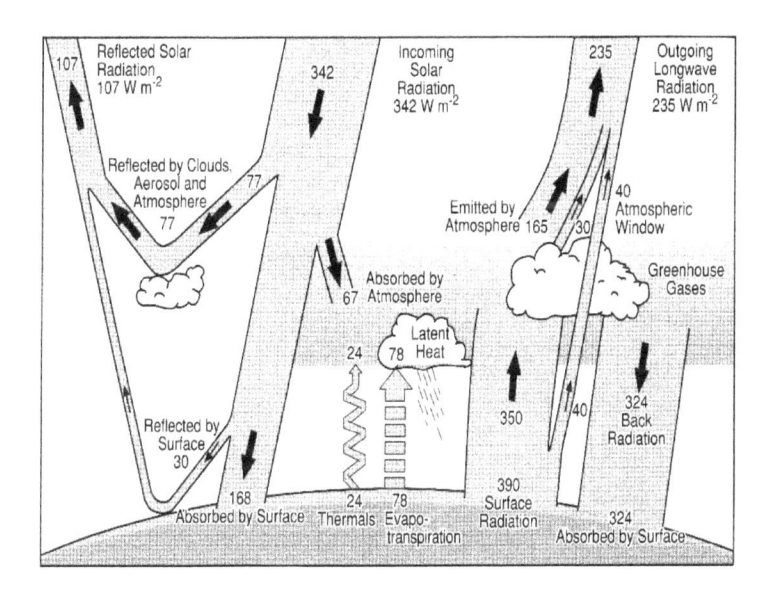

Figure 6.3
Earth's Annual Global Mean Energy Budget

It was described in detail by Kiehl and Trenberth (1997).[174]

The climate is considered to be balanced with the entering energy equal to the energy leaving. In practice this amounts to a belief that the whole climate system is in equilibrium. The whole object of this concept is to assume that any extra energy (radiative forcing) that can be detected by a rise on temperature is the results of increases in human induced greenhouse gases.

Heat input (radiative forcing) is assumed to be exclusively radiative, and thus to be instantaneous. It is measured in units of

---

[174] Kiehl, T., and Trenberth, Kevin E., 1997, *Earth's Annual Global Mean Energy Budget*, Bull. Amer. Meteor. Soc., 78, 197-208, doi: http://dx.doi.org/10.1175/1520-0477(1997)078<0197:EAGMEB>2.0.CO;2

Watts per square meter, W/m².

The model assumes that the earth can be considered to be flat with mean quantities providing input and output of energy. The apparent curvature of the earth in the diagram is illusory, as all quantities operate strictly up and down.

The heat engine that is the climate provides the energy for living organisms plus energy for erosion mountain building and changes in topology. The model ignores this and essentially treats the earth as dead and uninhabited. Yet somehow the dead earth succeeds in emitting greenhouse gases.

No part of the earth is ever in equilibrium or could possess an energy balance. There is excess energy in daytime and in the summer and a deficit at night and in the winter. The imbalance can change with every breath of wind, every cloud and every change of weather. Due to its vast amount of water, the earth has a large thermal capacity to absorb energy or lose it, for short or long periods.

The figures on the diagram turn out to be balanced but since they are mean figures there should be large uncertainties attached to each of them which destroy the balance. The calculated effects of greenhouse gases could not be distinguished from uncertainties in these figures.

The figure for the radiant energy emitted by the earth is given as 390 W/m².

If the earth is assumed to be black body then the radiation intensity from the earth, E, in W/m² is related to the absolute temperature T in K by the Stefan-Boltzmann equation:

$$\epsilon = \sigma T^4$$

Where σ is the Stefan-Boltzmann constant 5.67x 10^{-8} W/m².

Although there is no reliable evidence that this is so, the average temperature of the earth is thought to be 288K (15°C). If you put 288 into the Stefan-Boltzmann equation, you get 390

W/m^2, the amount shown in Figure 6.3. This means that they assumed that the earth is an ideal black body with a constant temperature of 288K (15°C). Such an assumption is absurd.

Trenberth et al[6] admit that there are Inter-annual variations in their atmospheric heat budget and they give details of local and temporal variability, dependent on changes in Sea Surface Temperature and ENSO events.[175] This variability, as well as the other errors, frequently exceed the supposed perturbations of greenhouse gas emissions.

The diagram and the models also assume other inappropriate averages, such as the supposed well-mixed concentrations of all the greenhouse gases, where distribution curves are actually necessary.

The 3rd IPCC Report (2007) had a slightly different diagram where several of the figures were changed, but there was still an unbelievable balance.

---

[175] Houghton, J. T., Ding, Y., Griggs, D.J., Noguer, M., van der Linden, P.J., Dai, X., Maskell, K., and Johnson, C.A., editors, 2001, *Climate Change 2001: The Scientific Basis*, Cambridge University Press, p. 90

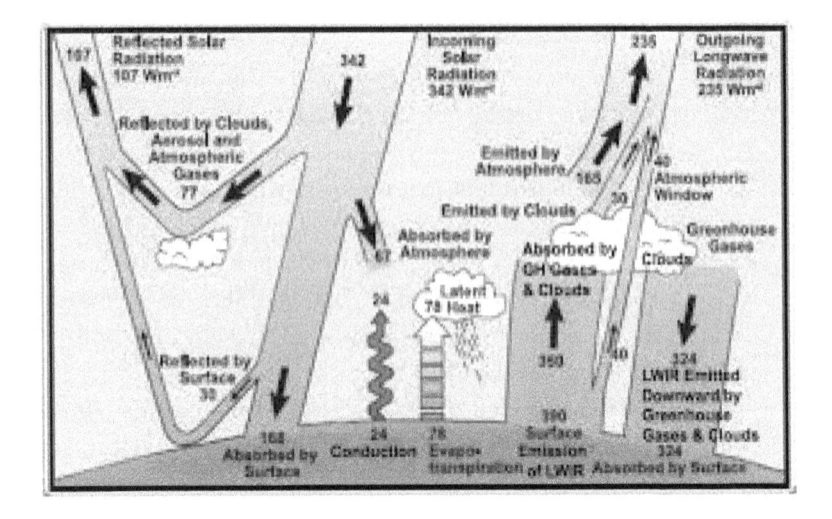

Figure 6.4
Earth's Mean Energy Budget in 3rd IPCC Report

Fasullo and Trenberth (2008) extended the model to cover changes in latitude and in seasons and the difference between land and ocean.[176] It still remains an unrealistic static concept unable to cope with the fact that half of the earth at any one time has no solar radiation at all, and that equilibrium and balance are never established.

Then, in a paper by Trenberth, Fasullo and Kiehl (2009), a complete revision was made.[177]

---

[176] Solomon, S., Qin, D., Manning, M.R., Marquis, M., Averyt, K., Tignor, M.H., Miller, H.L. and Chin, Z., editors, *Climate Change 2007: The Physical Science Basis (IPCC)*, Cambridge University Press, Chapter 1 FAQ 1.1. Fig 1

[177] Fasullo, J. T., Kiehl, T., Trenberth, K.E., 2008, *The Annual Cycle of the Energy Budget, Part I Global Mean and Land-Ocean Exchanges*, J Clim. 27, 2297-2311; Part II, *Meridional Structures and Poleward Transports*, 2313-23

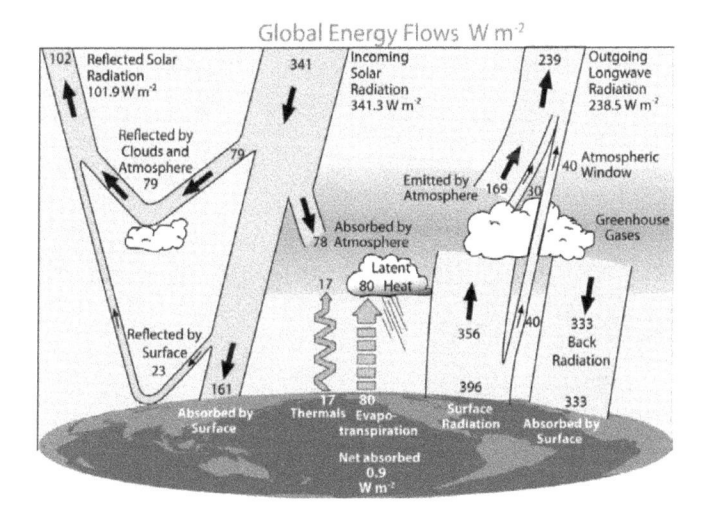

Figure 6.5
Revision by Trenberth, Fasullo & Keihl

This version represents the period from March 2000 to May 2004 and it caused something of a sensation when it first appeared, as for the first time there is no balance but a surplus of 9 W/m^2 which is much greater than the 1.6W/m^2 which is supposed to be the total effect of greenhouse gas emissions since 1750. The figures are all uncertain and variable and this surely means that they are incapable of telling whether increases in greenhouse gases affect the climate.

The latest version, in the 5th IPCC Report is shown in Figure 6.6.

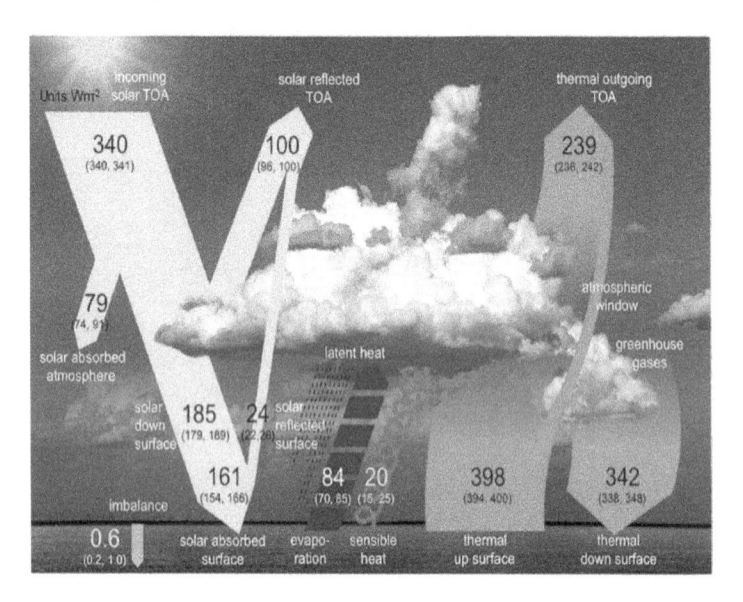

Figure 6.6
The Earth's Annual Global Mean Energy Budget

This version makes it a wholly obvious that the models cannot show the influence of greenhouse emissions. For the first time they have given uncertainty figures for each energy estimate.[178]

Radiative Forcing

The models are used to calculate the additional global input energy, radiative forcing, caused since the year 1750. Its breakdown is shown in Figure 6.6.[179]

It does not include the main greenhouse gas, water vapour, which is treated as a feedback. This is highly ridiculous since water vapour has a huge, largely unknown range of

---

[178] Ibid. IPCC, 2013, Chapter 2, Figure 2.11, p. 181
[179] Ibid. IPCC, 2013, *Summary for Policymakers*, p. 12

concentration and variability. It is just assumed to be related only to the assumed global temperature. Since carbon dioxide causes warming, water vapour increase causes enhancement of carbon dioxide warming. This is the tail wagging the dog.

Column water vapour can now be measured by AIRS satellites. Figure 6.7 shows an example.[180] Obtaining a global average from this diagram is impossible.

My earlier book pointed out the large range of assumptions for cloud feedback for the Third 1PCC Report.[181] Figure 6.6 shows that this has not changed. Model assumptions as a function of latitude were not very impressive (see Figure 6.9).[182]

Their current estimate of $2.3 W/m^2$ (1.33 to 3.33) for the radiative forcing caused by greenhouse gas increases since 1750 is well below the uncertainties of this model. The maximum forcing "expected" in 2100 of nearly $8 W/m^2$ is still smaller than the possible extra that could come from the model. It is therefore worthless as a means of estimating the influence of increases in greenhouse gases. It does not seem to be worth arguing about different values of "climate sensitivity" as they are all much less than the uncertainties.

---

[180] AIRS, http://airs.jpl.nasa.gov/resources/todays_earth_maps/water_vapor_total_column

[181] Gray, V. R., 2002, *The Greenhouse Delusion: A Critique of Climate Change 2001,* Multiscience Publishing, Essex, UK, p 38

[182] San Jose State University, http://www.sjsu.edu/faculty/watkins/watervapor01.htm

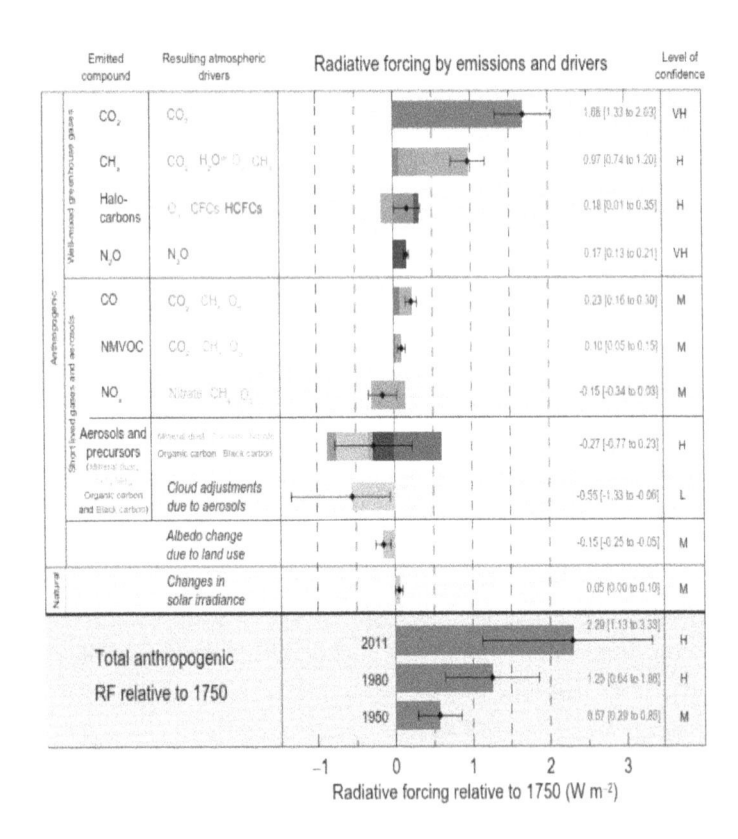

Figure 6.6
Components of Radiative Forcing Since 1750

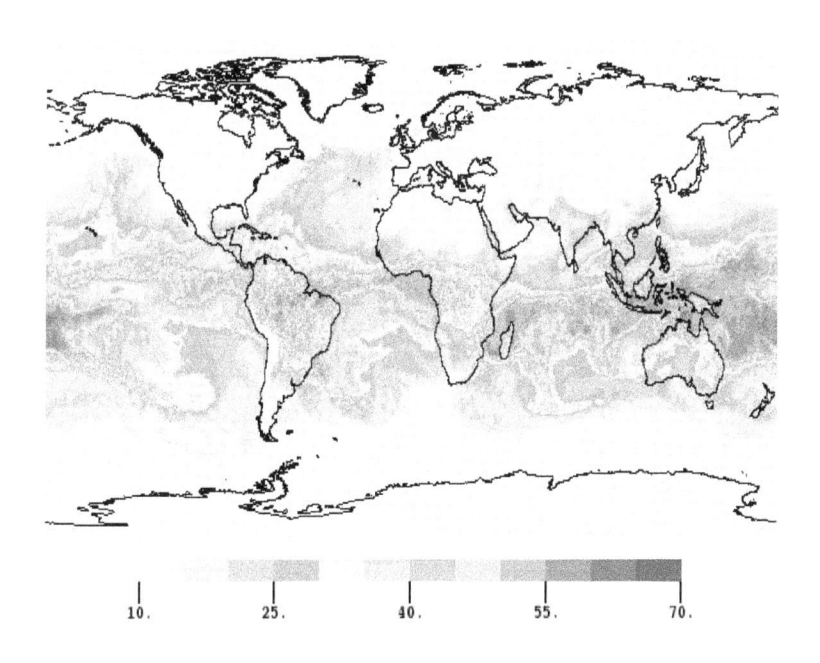

Figure 6.7
Column Water Vapour as Measured by AIRS

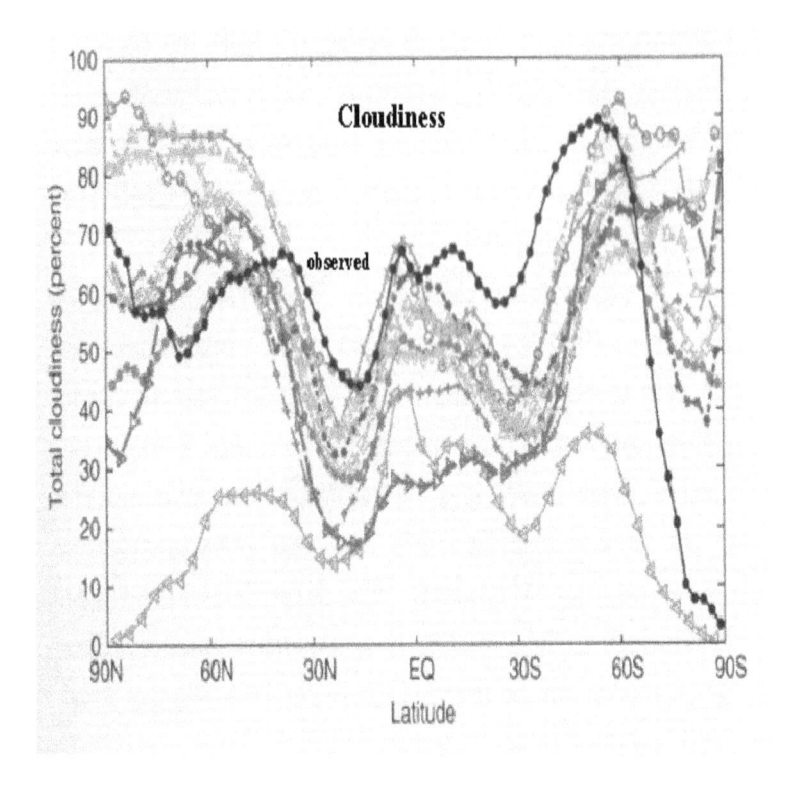

Figure 6.8
Model Assumptions of Cloudiness

Evaluation of Models

The IPCC does not make future predictions, only projections whose value depends on the assumptions of the models themselves and also on the scenarios which make a range of assumptions on how fast the additional emissions of greenhouse gases may happen in the future. These projections combine two different levels of uncertainty, those of the models and those of the scenarios.

Instead of trying to estimate the uncertainty of these

calculations by direct comparison between projections and actuality they carry out the following procedure:

> *...for AR5, the three IPCC Working Groups use two metrics to communicate the degree of certainty in key findings:*
>
> *Confidence is a qualitative measure of the validity of a finding, based on the type, amount, quality and consistency of evidence (e.g., data, mechanistic understanding, theory, models, expert judgment) and the degree of agreement; and (2) Likelihood provides a quantified measure of uncertainty in a finding expressed probabilistically (e.g., based on statistical analysis of observations or model results, or both, and expert judgement).*
>
> *Each key finding is based on an author team's evaluation of associated evidence and agreement. The confidence metric provides qualitative synthesis of an author team's judgement about the validity of a finding, as determined through evaluation of evidence and agreement. If uncertainties can be quantified probabilistically, an author team can characterize a finding using the calibrated likelihood language or a more precise presentation of probability. Unless otherwise indicated, high or very high confidence is associated with findings for which an author team has assigned a likelihood term.*[183]

---

[183] IPCC, 2013, p. 139, Para 1.4.1. 141 and Figure 1.11

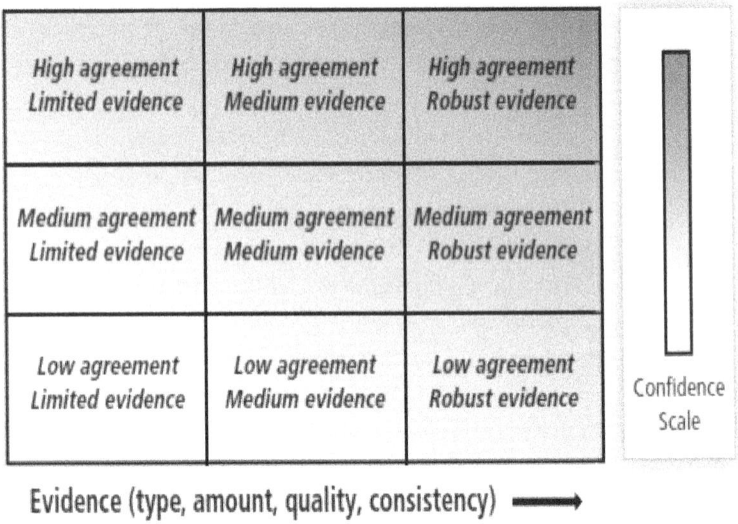

High agreement Limited evidence	High agreement Medium evidence	High agreement Robust evidence
Medium agreement Limited evidence	Medium agreement Medium evidence	Medium agreement Robust evidence
Low agreement Limited evidence	Low agreement Medium evidence	Low agreement Robust evidence

Evidence (type, amount, quality, consistency) ⟶

Confidence Scale

**Figure 6.9
Evaluating Models**

All these are nothing more than the personal opinions of the dragooned zombie scientists who are under the absolute control of the environmentally motivated government representatives who control the IPCC. This control is now absolute. It was displayed when the 1995 Final Draft was made to agree with the government dictated Summary for Policymakers and more recently when the Final Draft was not approved until it had been thoroughly checked for reliability. The penalty for independent thinking is instant dismissal.

The IPCC Scenarios

There have been three sets of scenarios. The first, launched with *Climate Change 90*, consisted of four scenarios, A, B, C, and D.[184] Scenario A was termed "Business As Usual" (BaU) or as SA90.

The next series of six scenarios was developed for the Supplementary Report *Climate Change 92* and the details are available from another supplementary Report. The scenarios were designated IS92a, IS92b, IS92c, IS92d, IS92e and IS92f.

I published a paper which showed that the scenarios were not plausible.[185] Its publication contributed to the dismissal not only of the Editor but of the entire editorial Board as shown in Chapter 8 on Climategate.

A new set of scenarios was used in *Climate Change 2001*. The scenarios were prepared by a special committee of IPCC Working Group III. Their Report, *Special Report on Emission Scenarios*, was produced without any input from the scientists involved with Working Group I, and its conclusions were foisted on *Climate Change 2001* without opportunity for discussion. Six teams of specialists from 18 countries drew up a total of 40 scenarios, all based on four "storylines" claimed to represent different views of what might happen in the future. The 40 scenarios were originally summarised in the form of four "Marker Scenarios," A1, A2, B1 and B2, but, as time went on, they split A1 into three: A1F1, A1B and A1T. These scenarios were also used for the Fourth IPCC Report 2007.[186]

In my previous book, I showed that these scenarios not only failed to predict the future, but in some respects failed to predict

---

[184] Ibid, *Climate Change : The IPCC Scientific Assessment*
[185] Pepper, W., et al, *Emissions Scenarios for the IPCC: An Update*, 1992
[186] Ibid.

the past.[187]

A series of Intercomparison exercises were carried out with the object of providing a set of approved models. These were the result of The Coupled Model Intercomparison Process (CMIP) which had achieved a third iteration by the 4th Report 2007. CMIP5 models were used for the calculations of the 5th Report.

A new approach to Scenarios was a feature of the Fifth Report which now has a set of Representative Concentration Pathways 1-5. Instead of a detailed history of emissions, each pathway was given a target reduction in radiative forcing by the year 2100. The four RCPs, RCP2.6, RCP4.5, RCP6, and RCP8.5 are named after a possible range of radiative forcing values in the year 2100 relative to pre-industrial values (+2.6, +4.5, +6.0, and +8.5 W/m^2, respectively.[188]

---

[187] Ibid. *The Greenhouse Delusion: A Critique of Climate Change 2001*

[188] Ibid. IPCC, 2013, *Technical Summary*, p. 83, Figure TFE6

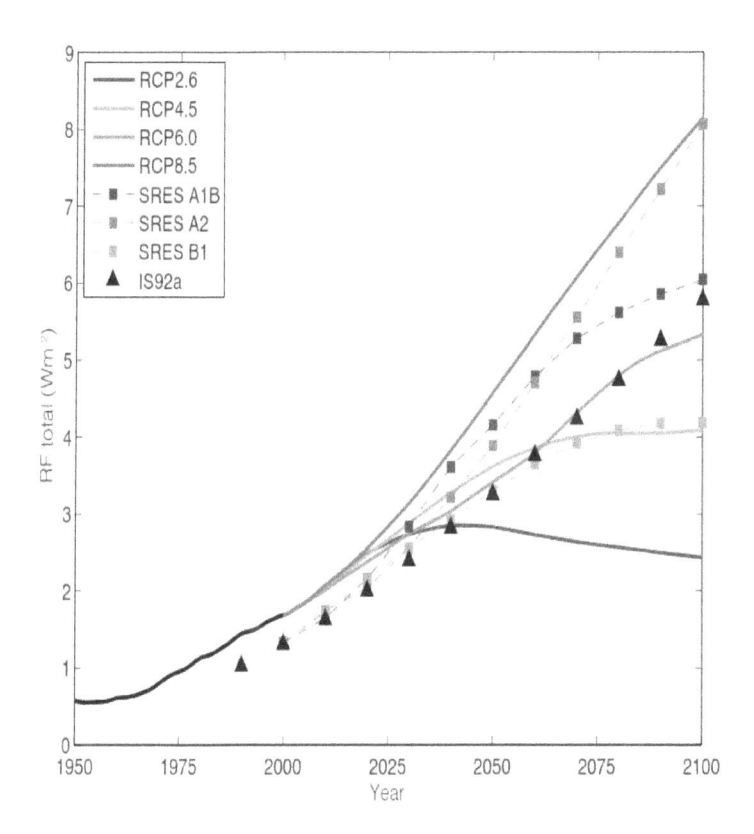

Figure 6.10
The Relationship between the Scenarios

Climate Sensitivity

Most scientists would agree that water vapour, carbon dioxide and other trace gases cause a warming of the global climate as a result of absorption of the infrared radiation from the earth in their spectral bands.

Weather forecasting meteorologists measure the many properties of the climate and provide a daily presentation of their influence on the global climate, but they have never found evidence that trace gas concentrations are sufficiently important in forecasting even to require regular measurement.

Scientists involved with the Intergovernmental Panel on Climate Change (IPCC) argue that carbon dioxide and other trace gases are not only important, but are the main cause of climate warming since 1750 and responsible for further warming as the concentrations rise.

The equilibrium climate sensitivity quantifies the response of the climate system to constant radiative forcing on multi-century time scales. It is defined as the change in global mean surface temperature at equilibrium that is caused by a doubling of the atmospheric CO_2 concentration.

It may be defined thus:

$$\Delta T_{CS} = \Delta T_T \times \Delta F_{CS} / \Delta F_T$$

Where:

ΔT_{CS} is the Climate Sensitivity

ΔT_T is the temperature change since 1750

ΔF_{CS} is the radiative forcing from doubling carbon dioxide

ΔF_T is the radiative forcing since 1750

The current IPCC range of values for equilibrium climate sensitivity are shown in Figure 6.9.

They regard a range between 1.5°C and 3.5° as likely, with high confidence, extremely unlikely below 1°C with high confidence and very unlikely above 6°C with medium confidence

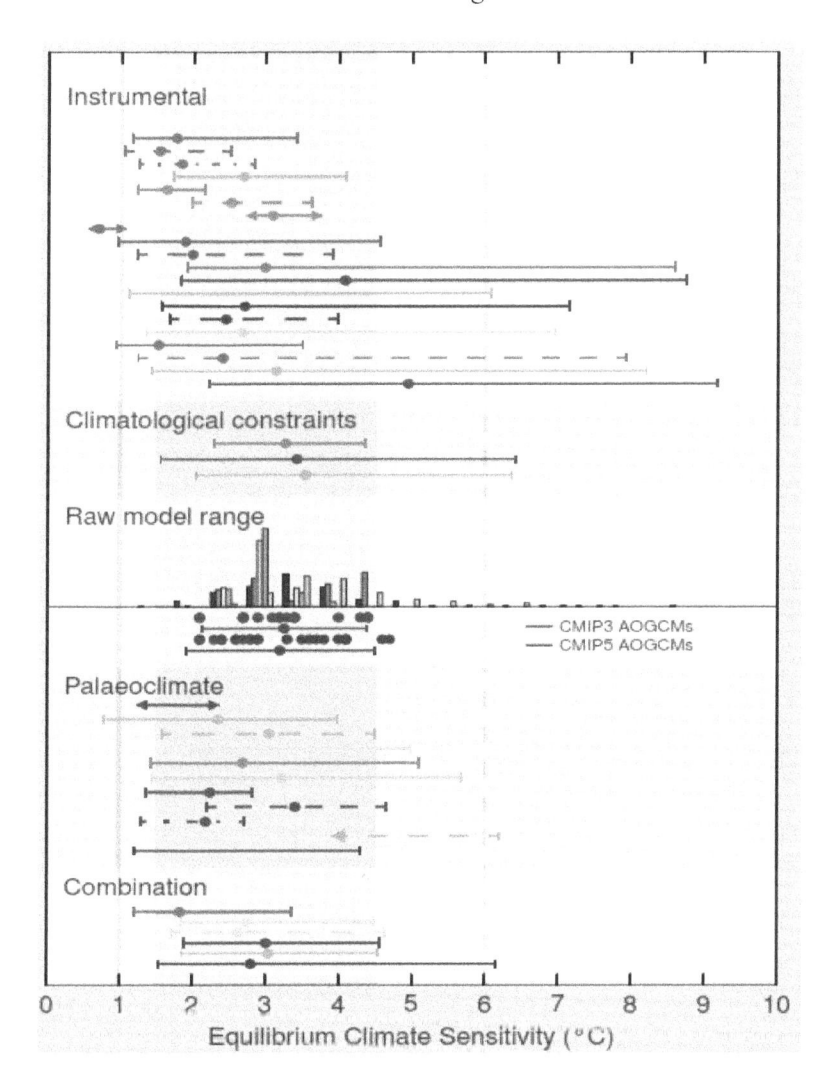

Figure 6.11
Current IPCC Figures for Equilibrium Climate Sensitivity

Annex II of The IPCC 5th Report[16] includes figures for CO_2 and other greenhouse gases from 1750 to 2011 in order to calculate the forcing from 1750 to 2000 and the forcing from 1850 to 2000 which is used to calculate climate sensitivity, using the assumption that all the temperature change from 1850 to 2000 was caused by greenhouse gases.

Geologists know that there are changes of climate for natural reasons in every geological period, short or long, whether or not human influences existed. There have been several where carbon dioxide concentrations were not related to assumed temperatures.

The FCCC assumption that all natural climate properties do not change from 1850 to 2000, but are merely variable, is unlikely to be true. Perhaps some or all of the claimed temperature change since 1850 had natural causes. The First IPCC Report (1990) suggested that recent temperature increases could have been a recovery from the Little Ice Age (1550-1850). Some increase was due to urban development and some to the persistent attempts to manipulate the record—as summarised by D'Aleo and Watts.[189]

The Mean Global Annual Surface Temperature Anomaly, whose origins are described in Chapter 7, is now incompatible with all the current models as is shown by this diagram shown in Figure 6.12, taken from the Technical Summary of AR5.[190]

---

[189] Nakicenovic, N., and Swart, R., editors, 2000, *IPCC Special Report: Emissions Scenarios*, Cambridge University Press.
[190] Ibid. *The Greenhouse Delusion: A Critique of 'Climate Change 2001'*, p. 67

Global mean temperature near-term projections relative to 1986-2005

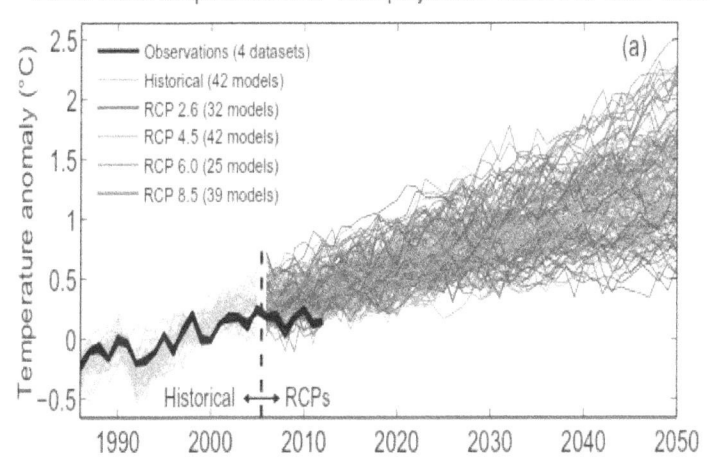

**Figure 6.12
Comparison between the IPCC Mean Annual Global Surface Temperature Anomaly and the Current IPCC Climate Models**

Because of this failure and the fact that the IPCC Mean Annual Global Temperature anomaly has not changed for the past 18 years they have decided to treat it on a decadal basis instead, as follows:

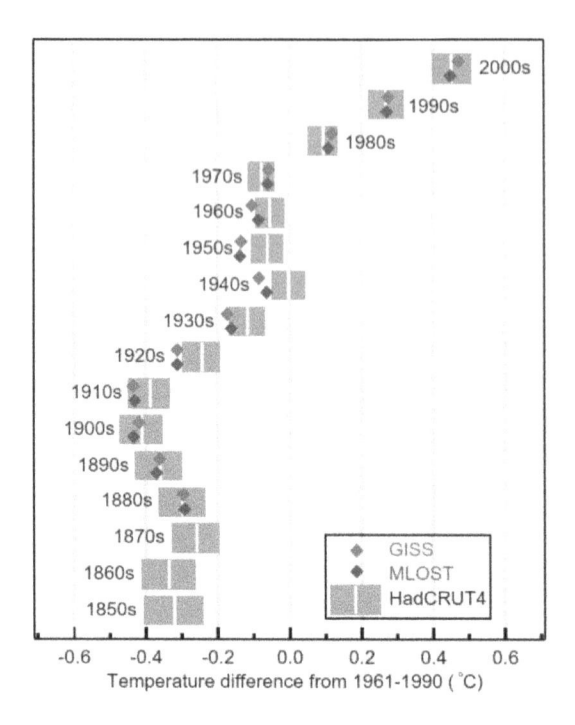

Figure 6.13
IPCC Decadal Temperatures

This irregularity is simply not compatible with a theory that temperature increases are caused by a steadily increasing concentrations of greenhouse gases.

There is a much more plausible temperature anomaly record from measurements in the lower atmosphere since 1978 by Microwave Sounder Units (MSU) on NASA satellites which measure the microwave spectrum of oxygen. Their results are confirmed by weather balloons which have been providing records since 1958. Since 2000, of the records both are beginning to resemble one another.

They provide further evidence that IPCC models are

currently incapable of predicting climate properties. These results suggest that IPCC figures for climate sensitivity are far too high, and the opinion of the meteorologists that its value is negligible is confirmed.[191]

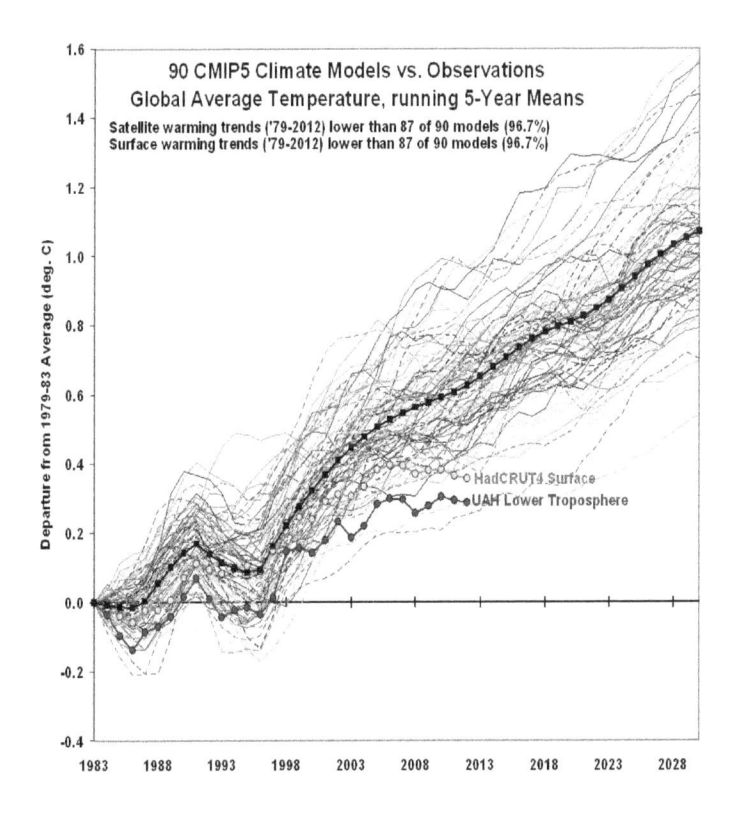

Figure 6.14
Comparison between Surface Temperature Records and IPCC Climate Models

---

[191] Ibid. IPCC 2013, Annex 1, Supplementaries

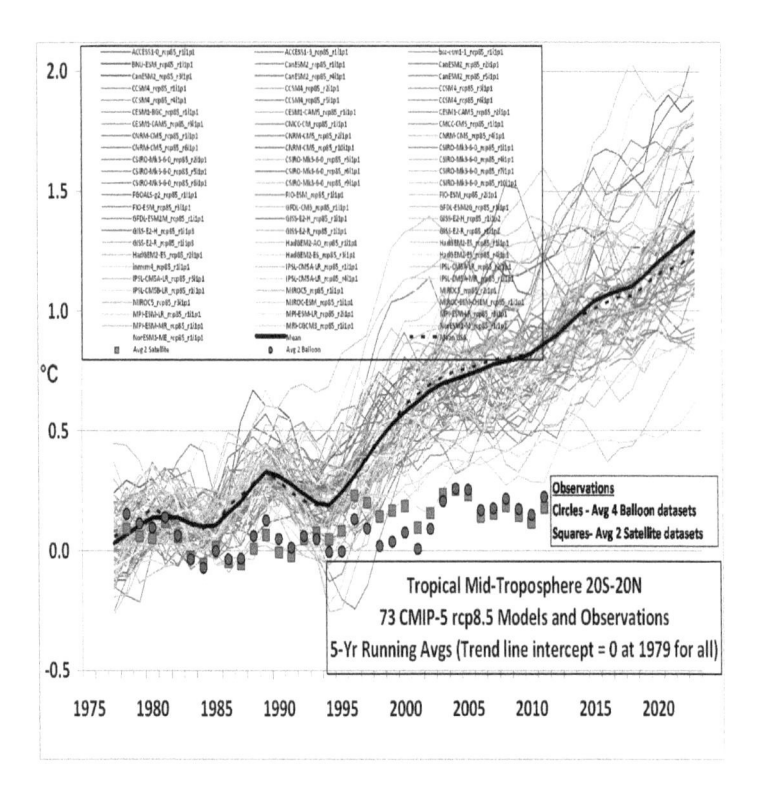

**Figure 6.15
Comparison between Temperature
Measurements of the Lower Atmosphere and
Surface with Climate Models**

The IPCC models project a hotspot. Figure 6.15 shows that this hotspot does not exist when compared with temperature observations by satellites and weather balloons.

The IPCC models rejected all of the measurements used by meteorologists except those used as feedbacks for their radiative forcing. The forcing is based on the belief that all temperature rise since 1750 was caused by greenhouse gas emissions, and it was calibrated from the temperature trend established by their own flawed and biased temperature anomaly record, which has now has become decadal because of the unexpected pause of the past 18 years.

They have now compiled projections to the year 2100 and map changes using all 42 of the CMIP5 models and all the RCP Scenarios for temperature and precipitation change for different seasons for some 21 different regions. These are presented in time series and global maps in Annex 1 and 4 Supplementary Reports to Annex 1. The temperature graphs all start with the models shown in Figure 6.15 above which makes them equally implausible with very large uncertainties.

They are quite frank about the adequacy of their models which they insist are not forecasts in some of the Supplementary Information to the Fifth Report.

> *Projections of future climate change are conditional on assumptions of climate forcing, affected by shortcomings of climate models and inevitably also subject to internal variability when considering specific periods. Projected patterns of climate change may differ from one climate model generation to the next due to improvements in models. Some model inadequacies are common to all models, but so are many patterns of change across successive generations of models, which gives some confidence in projections. The information presented is intended to be only a starting point for anyone interested in more detailed information on projections of future climate change and complements the assessment in*

Chapters 11, 12 and 14.

It is surely evident that the models have no predictive value and are therefore a complete failure. The estimates of climate sensitivity given by the IPCC are grossly exaggerated. The true figure is most probably near to that which is already assumed by the weather forecast meteorologists, negligibly low.

CHAPTER 7: THE GLOBAL WARMING SCAM

Temperature

TEMPERATURE IS THE degree of hotness of a substance. It may be thought of as the total mechanical and vibrational energy of the molecules of the substance.

The SI Unit is the Kelvin defined as the fraction 1/273.16 of the thermodynamic temperature of the triple point of water (exactly 0.01°C or 32.018°F). In other words, it is defined such that the triple point of water is exactly 273.16 K.

The most important climate property for establishing the theory that climate is influenced by so-called greenhouse gases is the global temperature which is considered to increase with increases in human induced emissions and therefore causing warming.

Unfortunately, a scientifically established global temperature does not exist, so it is impossible to know whether it is increasing or decreasing.[192]

---

[192] Essex, C. and McKitrick, R., 2008, *Taken by Storm*, Key Porter Books, Toronto, Canada

Vincent Gray

Essex et al explained this as follows:

> *Physical, mathematical and observational grounds are employed to show that there is no physically meaningful global temperature for the Earth in the context of the issue of global warming. While it is always possible to construct statistics for any given set of local temperature data, an infinite range of such statistics is mathematically permissible if physical principles provide no explicit basis for choosing among them. Distinct and equally valid statistical rules can and do show opposite trends when applied to the results of computations from physical models and real data in the atmosphere. A given temperature field can be interpreted as both "warming" and "cooling" simultaneously, making the concept of warming in the context of the issue of global warming physically ill-posed.*[193]

Temperature is an intensive property. It can only be defined as a property of a uniform substance.

> *Thermodynamic variables come in two varieties: extensive and intensive. Extensive variables are proportional to the size of the system. They are additive. In this category we find volume, mass, energy, entropy, particle number etc. We can combine two systems and the values of extensive variables for the whole system will simply be the sum of the values from the two components.*

---

[193] Essex, C., McKitrick, R. and Anderssen, B., 2007, *Does a Global Temperature Exist?*, J Non-Equilib. Thermodyn. 32, p. 1-27

Correspondingly a mean subsystem (loosely called the average) will have this sum divided by the number of components. Such an average over a quantity like mass is meaningful because the sum is meaningful. For example average mass is of importance to airlines because it is helpful to estimate the total load of an aircraft without having to weigh every passenger.

Intensive variables, by contrast, are independent of system size and represent a quality of the system: temperature, pressure, chemical potential etc. In this case combining two systems will not yield an overall intensive quantity equal to the sum of its components. For example two identical subsystems do not have a total temperature or pressure twice those of its components. A sum over intensive variables carries no physical meaning. Dividing meaningless totals by the number of components cannot reverse this outcome. In special circumstances averaging might approximate the equilibrium temperature after mixing, but this is irrelevant to the analysis of an out-of-equilibrium case like the Earth's climate.

It is possible to argue that sensors for temperature such as mercury in glass thermometers, thermocouples and thicknesses of tree rings or varves are surrogates or proxies of temperature. They can therefore be subject to statistical treatment, and even some form of averaging. However they are usually not representative samples, either of the surface or of the atmosphere, so the averages are of limited value and without scientific significance.

No point anywhere in the climate is ever in equilibrium. The only temperatures that exist are therefore transient infinitesimal increments of the surface and of the atmosphere which are constantly changing. There is never a single or an average temperature, only a transient temperature field. A physical model of the climate which assumes equilibrium can only be considered to be valid on a transient basis and can only be useful for forecasting if combined with records and weighed judgment of its likely future changes. This is essentially the procedure carried out for weather forecasting.

Geologists, climatologists and meteorologists have made attempts to derive rough estimates of trends in global surface temperatures from orbital and sun variability, fossil remains, tree rings, varves and historical evidence. These may be regarded as proxy temperatures. They are only partially based on actual scientific measurement and those that are used are often from unrepresentative samples. They have no scientifically plausible accuracy or bias estimates but they may represent collective opinion, often of self-styled experts, and are subject to the same differences of opinion as those shown amongst historians.

A comprehensive list of past proxy temperatures was compiled by Bernie Lewin.[194]

An example from 1975 is from Stanley.[195]

---

[194] Lewin, B.,
https://enthusiasmscepticismscience.wordpress.com/about/

[195] Stanley, S. M., 1989, *Earth and Life through Time*, p. 574, after J.M. Mitchell, in Energy and Climate; National Academy of Sciences, Washington, W.H. Freeman & Co

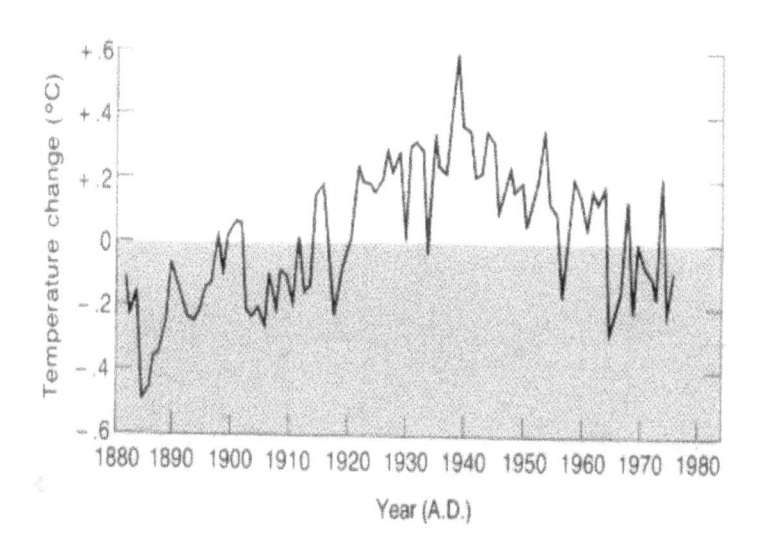

**Figure 7.1
Mean Northern Hemisphere Temperature Change from Weather Station Records, as Derived in 1975**

The 1990 Report of the Intergovernmental Panel on Climate Change gave a very useful summary of the then current opinions from a large number of references.[196]

---

[196] Houghton, .J T., Jenkins, G. J., and Ephraums, J.J., editors, 1990, *Climate Change : The IPCC Scientific Assessment*, Cambridge University Press

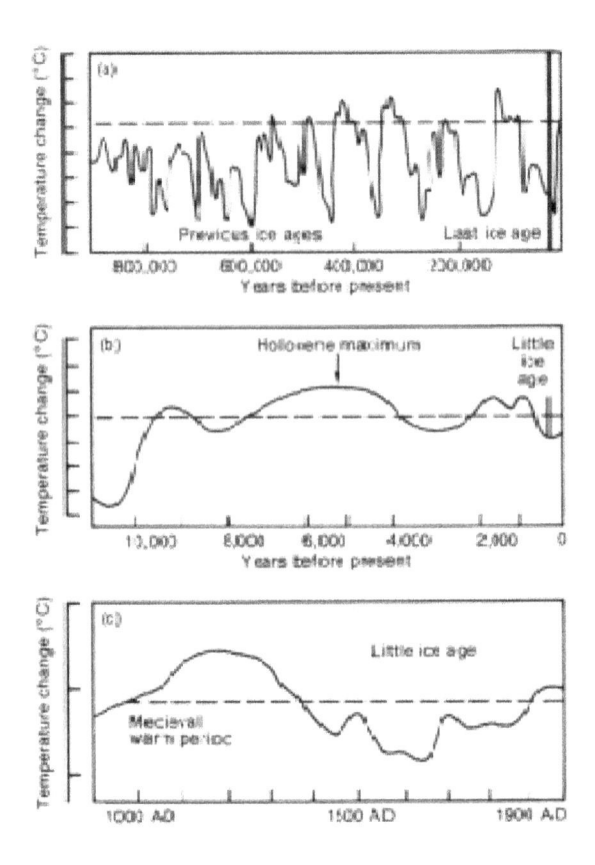

Figure 7. 2
Global Temperatures over Different Periods (IPCC 1994)

Harris and Mann provided the 2014 version from 2500 BC to 2040 AD as shown in Figure 7.3.[197]

---

[197] Harris, C., Mann, R., 2014, *Global Temperature Trends From 2500 B.C. To 2040 A.D*
http://www.longrangeweather.com/global_temperatures.ht5m

The Global Warming Scam

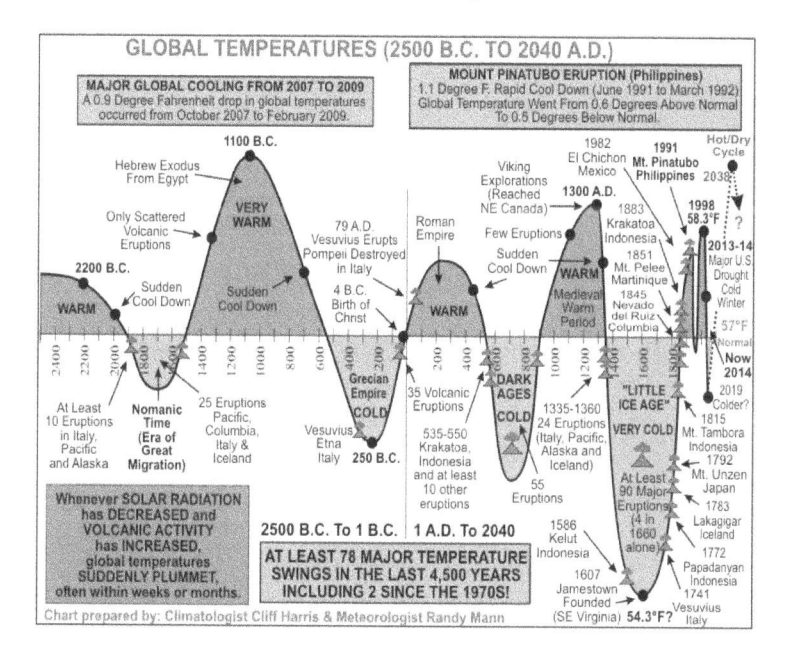

**Figure 7.3
Global Temperatures 2500 BC to 2040 AD**

It should be noted that they avoided mentioning any specific temperature values. The last two recognise the existence of a medieval warm period and a little ice age; the existence of both of which are now questioned by the IPCC.

Meteorologists realise that the temperature in any locality varies over a single day in a manner that cannot be predicted or averaged, but they have the task of forecasting future temperatures and other properties, so they trap a sample of the surface in a screen away from sunlight and buildings.

For some time they have measured the maximum and minimum temperature each day and used them to forecast future temperatures using computerised numerical models. These can only forecast with an accuracy of ±2°C with a bias of ±1°C for a week or so ahead. Nowadays many stations are able to measure hourly temperatures and frequently quote morning and afternoon figures besides the maximum and minimum.

The so-called Daily Mean Temperature—the mean of the maximum and minimum is not a genuine average but merely a guide to local range.

Intended as a student exercise, I show why no genuine local average surface temperature is possible by calculating the difference between a maximum/minimum average and an hourly average for a set of hourly temperature measurements for 24 weather stations in New Zealand,[198] for a typical winter's day (January 1st 2001) and a typical summer's day (July 1st 2001).[199]

For the winter series this difference varies from -0.2°C to +1.9°C for individual plots. For the summer series the difference varies from -1.1°C to +0.4°C.

The hourly averages are not symmetrical, so a daily average plus an estimate of accuracy such as a standard deviation is not possible. Here are some samples:

---

[198] Mackintosh, L., 2012, *Meteorologist for a day*, National Institute for Water and Air Research, New Zealand (NIWA),
http://www.niwa.co.nz/education-
andtraining/schools/resources/climate/meteorologist

[199] Gray, V. R., 2012, *Hourly Temperature Change*,
http://theclimatescepticsparty.blogspot.co.nz/2012/08/hourly-temperature-change.html

The Global Warming Scam

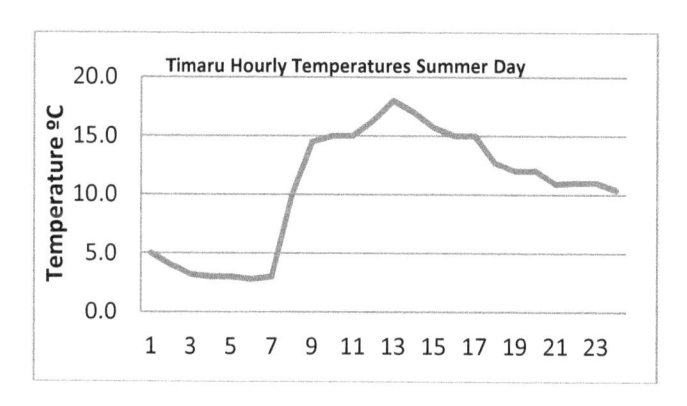

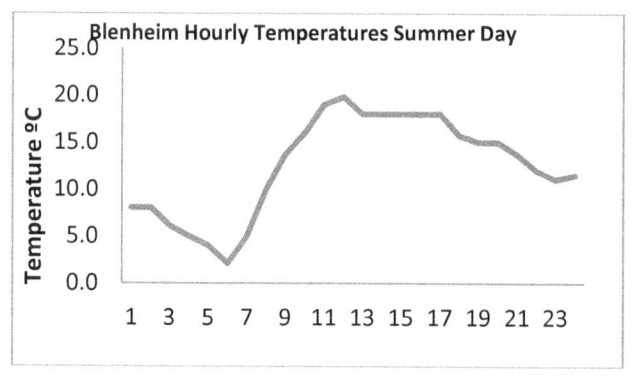

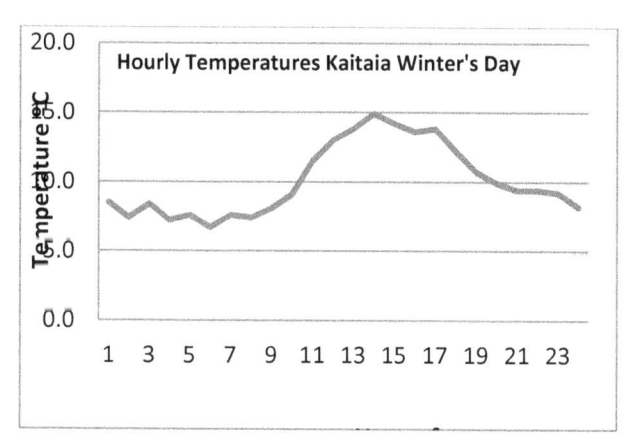

Timaru Hourly Temperatures Summer Day

Blenheim Hourly Temperatures Summer Day

Hourly Temperatures Kaitaia Winter's Day

215

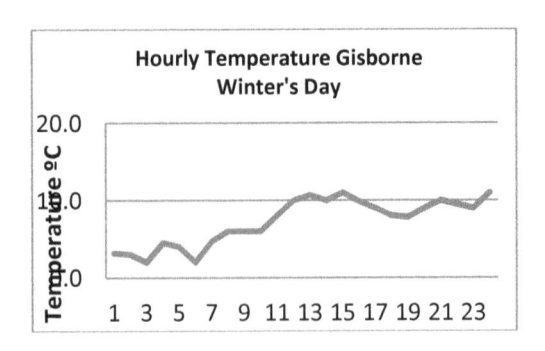

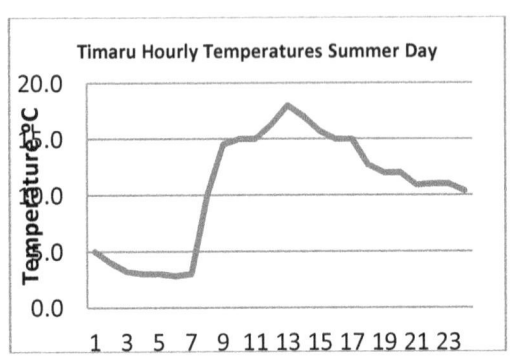

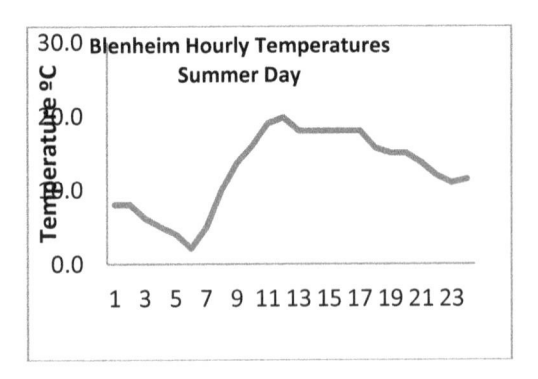

**Figures 7.4
Hourly Temperature Measurements**

The Global Warming Scam

Hansen's Initiative

On June 14 1988, Dr. James Hansen of the National Aeronautics and Space Administration told a U.S. Congress Committee that the globe was warming and that he had a method of measuring it.

On his website, he explains as follows:

> *The Elusive Absolute Surface Air Temperature (SAT)*
>
> Q. What exactly do we mean by SAT?
>
> A. I doubt that there is a general agreement how to answer this question. Even at the same location, the temperature near the ground may be very different from the temperature 5 ft above the ground and different again from 10 ft or 50 ft above the ground. Particularly in the presence of vegetation (say in a rain forest), the temperature above the vegetation may be very different from the temperature below the top of the vegetation. A reasonable suggestion might be to use the average temperature of the first 50 ft of air either above ground or above the top of the vegetation. To measure SAT we have to agree on what it is and, as far as I know, no such standard has been suggested or generally adopted. Even if the 50 ft standard were adopted, I cannot imagine that a weather station would build a 50 ft stack of thermometers to be able to find the true SAT at its location.
>
> Q. What do we mean by daily mean SAT?

A. Again, there is no universally accepted correct answer. Should we note the temperature every 6 hours and report the mean, should we do it every 2 hours, hourly, have a machine record it every second, or simply take the average of the highest and lowest temperature of the day? On some days the various methods may lead to drastically different results.

Q. What SAT do the local media report?

A. The media report the reading of 1 particular thermometer of a nearby weather station. This temperature may be very different from the true SAT even at that location and has certainly nothing to do with the true regional SAT. To measure the true regional SAT, we would have to use many 50 ft stacks of thermometers distributed evenly over the whole region, an obvious practical impossibility.

Q. If the reported SATs are not the true SATs, why are they still useful?

A. The reported temperature is truly meaningful only to a person who happens to visit the weather station at the precise moment when the reported temperature is measured, in other words, to nobody. However, in addition to the SAT the reports usually also mention whether the current temperature is unusually high or unusually low, how much it differs from the normal temperature, and that information (the

The Global Warming Scam

anomaly) is meaningful for the whole region. Also, if we hear a temperature (say 70°F), we instinctively translate it into hot or cold, but our translation key depends on the season and region, the same temperature may be 'hot' in winter and 'cold' in July, since by 'hot' we always mean 'hotter than normal', i.e. we all translate absolute temperatures automatically into anomalies whether we are aware of it or not.

Q. If SATs cannot be measured, how are SAT maps created?

A. This can only be done with the help of computer models, the same models that are used to create the daily weather forecasts. We may start out the model with the few observed data that are available and fill in the rest with guesses (also called extrapolations) and then let the model run long enough so that the initial guesses no longer matter, but not too long in order to avoid that the inaccuracies of the model become relevant. This may be done starting from conditions from many years, so that the average (called a 'climatology') hopefully represents a typical map for the particular month or day of the year.

Q. What do I do if I need absolute SATs, not anomalies?

A. In 99.9% of the cases you'll find that

> anomalies are exactly what you need, not absolute temperatures. In the remaining cases, you have to pick one of the available climatologies and add the anomalies (with respect to the proper base period) to it. For the global mean, the most trusted models produce a value of roughly 14°C, i.e. 57.2°F, but it may easily be anywhere between 56 and 58°F and regionally, let alone locally, the situation is even worse.[200]

Having agreed with me that there is no such thing as an absolute surface temperature. He then claims that it is possible to measure the rate in which this nonexistent quantity changes.

He shows how he does it in the following paragraph, which actually appears at the beginning of the argument.

> The GISTEMP analysis concerns only temperature anomalies, not absolute temperature. Temperature anomalies are computed relative to the base period 1951-1980. The reason to work with anomalies, rather than absolute temperature is that absolute temperature varies markedly in short distances, while monthly or annual temperature anomalies are representative of a much larger region. Indeed, we have shown (Hansen and Lebedeff, 1987) that temperature anomalies are strongly correlated out to distances of the order of 1000 km.[201]

---

[200] Hansen, J., http://data.giss.nasa.gov/gistemp/abs_temp.html
[201] Ibid.

Here he has changed his mind. Instead of absolute surface temperature being elusive, now it merely 'varies markedly in short distances.'

In reality, they do not exist at all. The temperature varies markedly at all times in the same place and therefore cannot legitimately be assumed constant. Hansen and Lebedeff have even gone one step further by choosing a monthly value made up of the daily average of the maximum and minimum as the constant temperature of each weather site.[202] These are not measured temperatures so they are called data.

This assigned temperature for each weather station is then assumed to apply also over a circle of 1,200 km or more radius. Then, they are said to be strongly correlated with the assigned temperature for the neighbouring weather station.

In the abstract they say:

> *The temperature changes at mid- and high latitude stations separated by less than* **1000 km are shown to be highly correlated; at low latitudes the correlation falls off more rapidly with distance for nearby station...*

and...

> *Error estimates are based in part on studies of how accurately the actual station distributions are able to reproduce temperature change in a global data set produced by a three-dimensional general circulation model with realistic*

---

[202] Hansen, J., Lebedeff, S., *Global Trends of Measured Surface Air Temperature*, J Geophys Res, VOL. 92, No. D11, 13,345-13,372

variability.[203]

At their 1200 km distance, the highly correlated coefficient is around 0.5 and it gets smaller at lower latitudes. This is not a high enough figure to justify an assumption of good correlation. They cover up this unjustified assumption with models where the presumed correlation can be assumed higher.

The whole world is divided into 80 equal area boxes and each box into 100 sub boxes. These cover most of the surface, including the oceans, by including islands whose influence is a radius of 1200km. A monthly and then an annual average of each sub box is averaged to give one for the whole box and that average is subtracted from a global mean to give the annual temperature anomaly. The huge uncertainties involved in each of these steps cannot be estimated and are the ignored.

The number and length of records and cover is variable. The figures below indicate the number of stations with record length at least N years as a function of N, the number of reporting stations as a function of time, the percent of hemispheric area located within 1200km of a reporting station.

---

[203] Ibid.

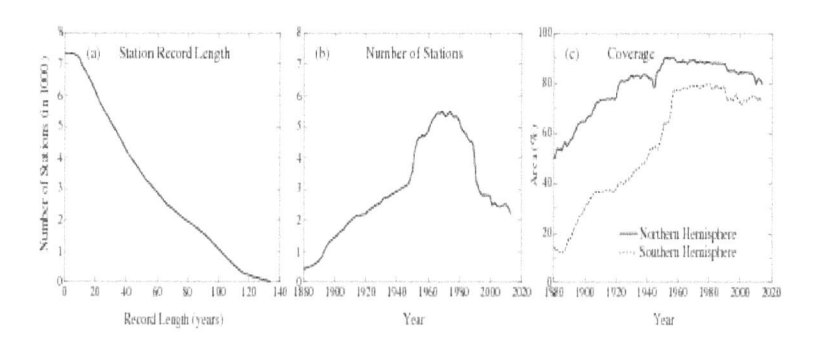

Figure 7.5
Variability of Temperature Records with Length, Number and Hemisphere

The anomaly record is not uniform. Standards of measurement vary from place to place and over time. Numbers and location of acceptable stations vary and they are not distributed uniformly over the earth. Whole regions may be omitted, particularly early ones.

The whole argument depends on correlation coefficients of as low as 0.5 or less for temperature anomaly differences. The resulting plots should have much larger uncertainty estimates.

Sea Surface Temperature

Hansen's system includes only land-based measurements. An average global temperature anomaly needs to include the 71% of the earth's surface that is ocean. There are many temperature measurements made from ships, but the quality control is much worse than on the land and even then, whole regions have no figures. Folland and Parker have claimed to have found a way of incorporating the data.[204] One difficulty is that many early measurements were from buckets drawn from the sea and it is sometimes uncertain whether the buckets were metal or wood. During the First World War, measurements could not be made on deck. Also, some measurements are from a weather station on board, often beneath the funnel.

Both American temperature compilers, the Goddard Institute of Space Studies (GISS) and The Global Historical Climatology Network (GHCN) have never accepted the use of the sea surface measurements for a global average and as a result they have to resort to a whole host of dubious devices to claim that their figures are "global." They use recent satellite measurements for the ocean and extrapolate them into the past.[205] It is very suspicious that incorporating the sea surface measurements seems to make little difference to either system.

Although only 29% of the earth's surface is land, the peak number of 5° x 5° grid-boxes from land-based weather stations were 880 in 1980, which includes 34% of the earth's surface. This has been achieved because there are many stations on small islands surrounded by ocean where the land temperature is assumed to be typical of the surrounding ocean. Also, a proportion of sea surface measurements are from fixed buoys and weather ships. These and the land stations measure above the surface, whereas current sea surface measurements are made from ship's engine intake, which is below the surface.

The IPCC has manipulated this unreliable average from weather stations and sea surface readings to provide the Mean

---

[204] Folland, C. K. and Parker, D. E., 1995, *Correction of Instrumental Biases in Historical Sea Surface Temperature Data*, Quart. J. Met. Soc. 15 1195-1218

[205] Reynolds, R.W., Rayner, N. A., Smith, T. M., Stokes, D. C. and Wang, W., 2002, *An Improved In Situ and Satellite SST Analysis for Climate*, J. Climate 15, 1609-1624

Global Surface Temperature Anomaly Record (MGSTAR, see Figure 7.6) which they use as a proxy for global temperature. Estimates of error are rare and have to ignore *unknown unknowns*.[206]

No point on the curve is an actual temperature. They are supposed to be a record of temperature anomalies, not a temperature record, and each point is the result of a manipulation process from a different number of unrepresentative and unstandardized readings. No plausible estimate of accuracy is possible, but it must be greater than the ±2°C and ±1°C currently possible from weather forecasts, so trends of a few tenths of a degree should be regarded as without statistical significance.

Like all proxies, its value depends on the extent to which recognised experts and the general public are prepared to accept it. It cannot be validated by any scientific procedure and the IPCC does not claim that it can be. It cannot provide forecasts, but only projections which are evaluated by scientists who depend for their income and career on the IPCC.

---

[206] Brohan, P., Kennedy, J.J., Harris, I., Tett, S. F. B. and Jones, P. D., 2006, *Uncertainty Estimates in Regional and Global Observed Temperature Changes: A New Data Set from 1850*, J. Geophys. Res. 111, D12106.doi:1020/2005JD006546

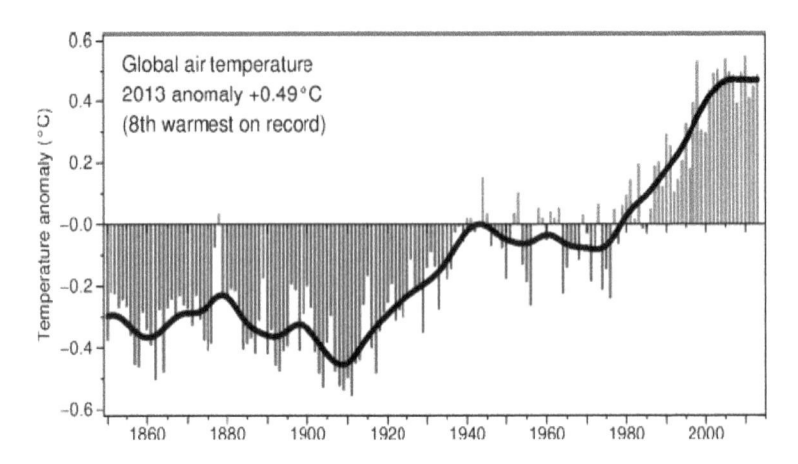

Figure 7.6
Global Air Temperature Anomaly 2013 1901-2000 Reference Period

Figure 7.6 shows the current version of the Hadley CRUT MGSTAR which finds a temperature anomaly of a mere 0.49°C since 1850.[207]

Frank confirmed that the supposed rise is statistically indistinguishable from zero.[208]

---

[207] Climatic Research Unit, 2014, http://www.cru.uea.ac.uk/
[208] Frank, P., 2010, *Imposed and Neglected Uncertainty in the Global Average Surface Air Temperature Index*, Energy and Environment, 21, Number 8 / December 2010 DOI:

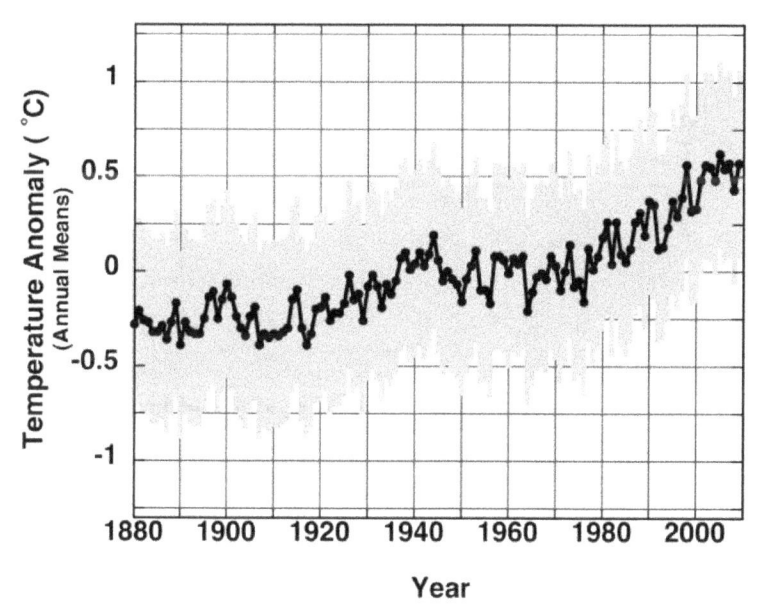

**Figure 7.7
Estimated 95% Uncertainties in the Mean
Annual Global Temperature Anomaly Record
(MAGTAR)**

Microwave Sensing Units

A similar proxy sequence has been derived from Microwave Sounding Units, satellite measurements of the microwave spectrum of oxygen in various levels of the atmosphere (MSU). The microwave spectrum of oxygen is dependent on temperature, so an average proxy temperature can be calculated for the various regions of the atmosphere. The segment of the upper troposphere represents global trends from 1958.

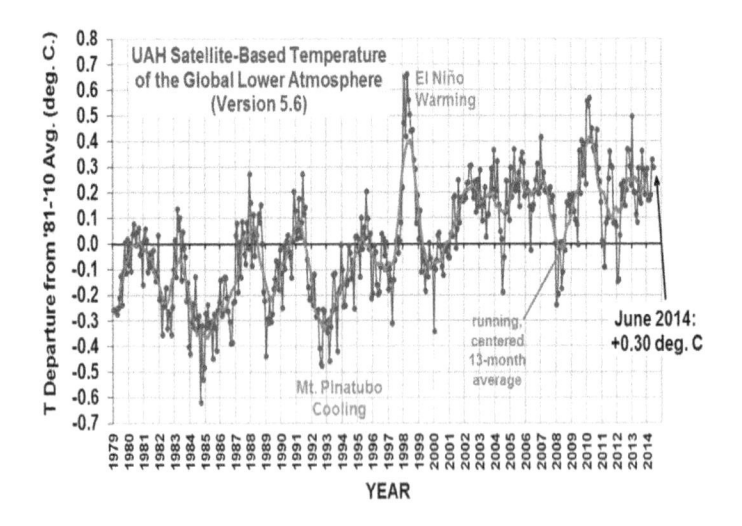

**Figure 7.8
Temperature Anomalies in the Upper Troposphere Since 1970**

The technique was developed by John Christy and Roy Spencer at the University of Alabama. The series in the Lower Troposphere (TLT) is presented as a comparison with the MGSTAR surface temperature record and is also given as anomalies where the entire series is the reference period. It has been running since 1979.[209]

At first this record appeared to differ markedly from the surface anomaly record preferred by the IPCC and there were long arguments as to which were authentic. Christy and Spencer were not supporters of the IPCC dogmas so a rival MSU Group was set up (Remote Sensing Systems, RSS) headed by IPCC

---

[209] Spencer, R., UAH update, http://www.drroyspencer.com/2014/04/uah-global-temperature-update-for-march-2014-0-17-deg-c-again/

sympathisers in the hope that it might prove to support the surface series. Now, the surface records and the rival MSU records possess considerable agreement and all of them show the absence of warming for the past 18 years which prove that the IPCC models are wrong.

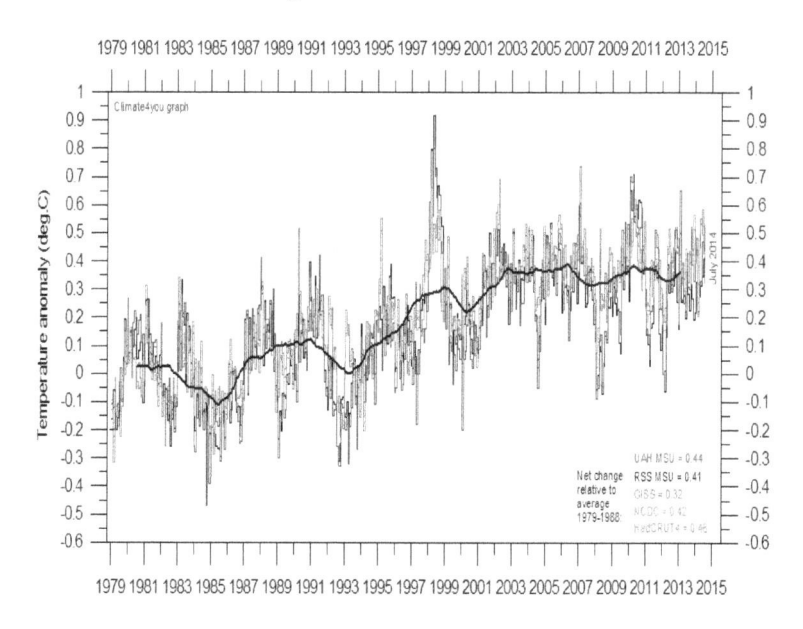

Figure 7.9
Comparisons between Proxy Temperature Anomaly Records

The IPCC has responded by abandoning its previous favourite and replacing it with Mean Decadal Surface Temperature Record (MDSTAR), as follows.[210]

---

[210] *Comparing global Surface Temperature Estimates*, http://www.climate4you.com/GlobalTemperatures.htm#

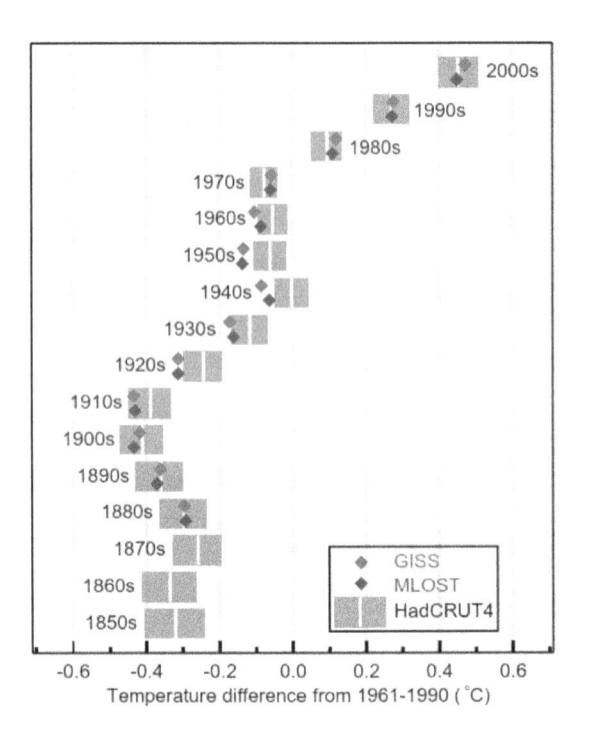

**Figure 7.9
Changes in Decadal Values of the Mean Global Decadal Temperature Record**

Even this system shows only one degree rise over 150 years, and two periods (870s to 1900s and 1940s to 1970s) when decadal temperatures fell.

It is amusing to examine the RSS versions of the MSU system which now quotes by the decade.[211]

---

[211] Remote Sensing Systems, http://www.remss.com/measurements/upper-air-temperature

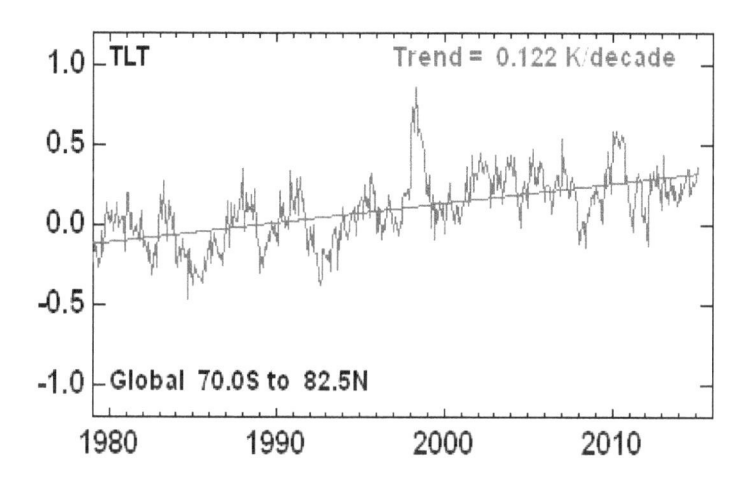

Figure 7.10
RSS MSU Temperature Anomalies

They have the foolishness to regard a temperature rise of only one tenth of degree with three places of decimals for a whole decade as having statistical significance.[212]

Fiddling the Figures

The MGASTAR has been the object of various doubtfully justified attempts to make its trend greater over the years with little apparent success. This has led the IPCC to argue that a trend should be calculated from decadal rather than annual

---

[212] IPCC, 2013, *Climate Change 2013: The Physical Science Basis*, Contribution of Working Group I to the Fifth Assessment Report of the Intergovernmental Panel on Climate Change, Stocker, T.F., Qin, D., Plattner, G-K, Tignor, M., Allen, S. K., Boschung, J., Nauels, A., Xia, Y., Bex, V., and Midgley, P.M.

temperature anomalies. If this does not work we may have to wait for whole centuries before we can become confident that the globe is really warming.

The record is very difficult to reconcile with the IPCC theory that there is a steady warming of the surface from increasing emissions of so-called greenhouse gases. The tiny claimed rise is easily explained by a modest overall upwards bias which would be expected from the increased population and urbanisation of the world since1850.

I pointed this out in 2000.[213] There are many records of urban influences on local temperature and there are a number of long records from stations that have shown little change which show no temperature rise. I also pointed out that liquid in glass thermometers automatically read high if they are not regularly calibrated, particularly with old ones, and there are many other inaccuracies which seem to be ignored.

Michaels and McKitrick have shown an important influence of sociological change and also from inhomogenities.[214,215]

There have been repeated attempts to boost the supposed rise in the MGASTAR by adjustments. It began right at the beginning with the attempts to argue that the record is not influenced by urban heating.

In 1990, two papers on the subject appeared in the two most respectable scientific journals. The first was by Phil Jones

---

[213] Gray, V. R., 2000, *The Cause of Global Warming*, Energy and Environment. 11, 613-629

[214] McKitrick, R.R. and Michaels, P.J., 2006, *A Test of Corrections for Extraneous Signals in Gridded Surface Temperature Data*, Clim. Res, 26, 150-173

[215] McKitrick, R.R. and Michaels, P.J., 2007, *Quantifying the Influence of Anthropogenic Surface Processes and Inhomogeneities on Gridded Global Climate Data*, J. Geophys. Res. 112, D24S09, doi:10:1029/2007JD008465

et al appearing in *Nature*.[216] The IPCC repeatedly quoted this paper by as evidence that urban heating is negligible.

These authors examined an extensive set of rural station temperature data for three regions of the world—European parts of the Soviet Union, Western Australia and Eastern China. When combined with similar analyses for the contiguous United States, the results are claimed to be representative of 20% of the land area of the Northern Hemisphere and 10% of the Southern Hemisphere.

They worked out the linear slope of temperature anomalies for the rural series in each case and compared it with the same slope for several gridded series. For the Western USSR, it covered the period 1901-1987 and 1930-1987, for Eastern Australia it was 1930-1988 compared with 1930-1997, for Eastern China it was 1954-1983 and for the contiguous United States it was 1901-1984. The differences between urban and rural slopes were only significant at the 5% level for Eastern Australia and for one set of Eastern China.

They concluded:

> *It is unlikely that the remaining unsampled areas of the developing countries in tropical climates, or other highly populated parts of Europe, could significantly increase the overall urban bias above 0.05°C during the twentieth century.*

---

[216] Jones, P. D., Ya, Groisman, Coughlan, P.M., Plummer, N., Wang, W.C. and Karl, T.R., 1990, *Assessment of Urbanization Effects in Time Series of Surface Air Temperature Over Land*, Nature 347, pps. 169-172

It is unclear whether this small correction has been made for the most recent version of the Jones et al global temperature series (see Figure7.6).

There are several things wrong with the Jones et al (1990) paper.

The quality of the data is even worse than usual. They admit that it is unfortunate that separate maximum and minimum temperature data are not more widely available.

The qualification for a rural site is a population below 10,000 for Western Soviet Union, below 35,000 for Eastern Australia, and below 100,000 for Eastern China. There is ample evidence (Ibid, Gray 2000) that urban effects exist in such places.

They have chosen countries with a continuous record of effective scientific supervision. These are not representative of the rest of the world, where changes of country and adequate supervision are far less common.

Even these countries raise doubts. Russia had a tyrannical regime where statistics were frequently manipulated for political purposes. China had a major famine from the *Great Leap Forward* between 1958 and 1959 and also a manipulation of statistics.

In the very same year there appeared in Geophysical Research Letters another paper which included two of the authors of the previous paper, Wang and Karl.[217]

The abstract of this paper reads:

> *We used 1954-1983 surface temperature from 42 Chinese urban (average population 1.7 million) and rural (average population 150,000) station pairs to study the urban heat island effects. Despite the fact that the rural*

---

[217] Wang, W-C, Zeng, Z., Karl, T.R., 1990, *Urban Heat Islands in China*, Geophys. Res. Lett. 17, 2377-2380

> stations are not true rural stations, the magnitude of the heat islands was calculated to average 0.23°C over the thirty year period, with a minimum value (0.19°C) during the 1964-1973 decade and maximum (0.28°C) during the most recent decades.

This study appears to have used the same stations that were claimed to have no urban bias in the first paper and now there is an urban bias even if *rural* now includes places with population as high as 150,000.

The early paper states of Eastern China:

> The stations were selected on the basis of station history: We chose those with few, if any, changes in instrumentation, location or observation times.

Wang et al says:

> They were chosen based on station histories. We chose those without any changes in instrumentation, location, or observation times.[218]

Both papers were written at the same time and different conclusions made from the same data. Douglas Keenan found that they both used the same data and he showed that some places did not exist and many of the Chinese stations moved

---

[218] Ibid.

several times over the period in question, in one case 15km.[219] He accused Wang of outright fraud.

Although Wang was cleared of this charge by his college, Tom Wigley, in this exchange from the Climategate papers, was in no doubt:

> From: Tom Wigley
> To: Phil Jones
> Subject: [Fwd: CCNet Xtra: Climate Science Fraud at Albany University]
> Date: Mon, 04 May 2009 01:37:07 -0600
> Cc: Ben Santer
>
> Phil,
> Do you know where this stands? The key things from the Peiser items are...
> Wang had been claiming the existence of such exonerating documents for nearly a year, but he has not been able to produce them. Additionally, there was a report published in 1991 (with a second version in 1997) explicitly stating that no such documents exist. Moreover, the report was published as part of the Department of Energy Carbon Dioxide Research Program, and Wang was the Chief Scientist of that program.

and...

> Wang had a co-worker in Britain. In Britain, the Freedom of Information Act requires that

---

[219] Keenan, D., 2007, *The Fraud Allegation Against Some Climatic Research of Wei-Chyug Wang*, Energy and Environment, 18, 985-995.

> data from publicly-funded research be made available. I was able to get the data by requiring Wang's co-worker to release it, under British law. It was only then that I was able to confirm that Wang had committed fraud.[220]

Despite this plain evidence of fraud, the paper by Jones et al[221] was still quoted in the next IPCC Report as evidence against urban heating, and it was even endorsed by reference in the most recent IPCC Report.[222]

D'Aleo and Watts documented many of the devices that have been used to try and boost the alleged warming shown by the MGASTAR.[223] Their Summary for policymakers shows what they have found:

> 1. Instrumental temperature data for the pre-satellite era (1850-1980) have been so widely, systematically, and unidirectionally tampered with that it cannot be credibly asserted that there has been any significant "global warming" in the 20th century.
>
> 2. All terrestrial surface-temperature databases

---

[220] *Reclaiming Climate Science*, http://www.greenworldtrust.org.uk/Science/Social/FOIA1241415427.txt.htm

[221] Ibid. *Assessment of Urbanization Effects in Time Series of Surface Air Temperature Over Land*

[222] Ibid. IPCC 2013

[223] D'Aleo, J. and Watts, A., *2010 Surface Temperature Records Policy-Driven Deception*
http://scienceandpublicpolicy.org/originals/policy_driven_deception.html

exhibit signs of urban heat pollution and post measurement adjustments that render them unreliable for determining accurate long-term temperature trends.

3. All of the problems have skewed the data so as greatly to overstate observed warming both regionally and globally.

4. Global terrestrial temperature data are compromised because more than three quarters of the 6,000 stations that once reported are no longer being used in data trend analyses.

5. There has been a significant increase in the number of missing months with 40% of the GHCN stations reporting at least one missing month. This requires infilling which adds to the uncertainty and possible error.

6. Contamination by urbanization, changes in land use, improper siting, and inadequately calibrated instrument upgrades further increases uncertainty.

7. Numerous peer-reviewed papers in recent years have shown the overstatement of observed longer term warming is 30-50% from heat-island and land use change contamination.

8. An increase in the percentage of compromised stations with interpolation to vacant data grids may make the warming bias greater than 50% of 20th-century warming.

9. In the oceans, data are missing and uncertainties are substantial. Changes in data sets introduced a step warming in 2009.

10. Satellite temperature monitoring has provided an alternative to terrestrial stations in compiling the global lower-troposphere temperature record. Their findings are increasingly diverging from the station-based constructions in a manner consistent with evidence of a warm bias in the surface temperature record.

11. Additional adjustments are made to the data which result in an increasing apparent trend. In many cases, adjustments do this by cooling off the early record.

12. Changes have been made to alter the historical record to mask cyclical changes that could be readily explained by natural factors like multi-decadal ocean and solar changes.

13. Due to recently increasing frequency of eschewing rural stations and favoring urban airports as the primary temperature data sources, global terrestrial temperature databases are thus seriously flawed and can no longer be representative of both urban and rural environments. The resulting data are therefore problematic when used to assess climate trends or VALIDATE model forecasts.

> 14. *An inclusive external assessment is essential of the surface temperature record of CRU GISS and NCDC chaired and panelled by mutually agreed to climate scientists who do not have a vested interest in the outcome of the evaluations.*
> 15. *Reliance on the global data by both the UNIPCC and the US GCRP/CCSP should trigger a review of these documents assessing the base uncertainty of forecasts and policy language.*

The examples given in this report can be supplemented by more recent ones. E. M. Smith (Chiefio) found that the hottest country in the world was Bolivia.[224] When he tried to access the figures he found there had not been any for several years. It seems there is an automatic adjustment that replaces a missing sequence with the average of nearby countries. As these were mainly coastal resorts in Mexico, they were much warmer. Then there was an automatic increase to allow for the presumed cold in Bolivia, so we end up with a high value.

I provided several other examples of how temperatures are being manipulated.[225]

Paul Homewood has some recent examples.[226] [227] Christopher Booker has found recent South American examples

---

[224] Chiefio, https://chiefio.wordpress.com/2010/01/08/ghcn-gistemp-interactions-the-bolivia-effect/
[225] Gray, V. R., 2008, *The Global Warming Scam*, http://www.climatescience.org.nz/images/PDFs/GlobalScam3a.pdf
[226] Paul Homewood, https://notalotofpeopleknowthat.wordpress.com/
[227] http://www.telegraph.co.uk/news/earth/environment/globalwarming/11395516/The-fiddling-with-temperature-data-is-the-biggest-science-scandal-ever.html

where the figures have been changed.[228] Jennifer Marohasy finds that Australia's Australian Bureau of Meteorology is also fiddling their figures.[229]

United States Temperatures

In 1999 Hansen et al gave the graph shown in Figure 7.11 for the U.S. temperature record.[230]

---

[228] Christopher Booker, http://www.telegraph.co.uk/comment/11367272/Climategate-the-sequel-How-we-are-STILL-being-tricked-with-flawed data on global warming.html
[229] Jennifer Marohasy, http://jennifermarohasy.com/2014/08/heat-is-on-over-weather-bureau-homogenising-temperature-records/
[230] Hansen, J., Ruedy, R., Glascoe, J. and Satom, M., August 1991, http://www.giss.nasa.gov/research/briefs/hansen_07/

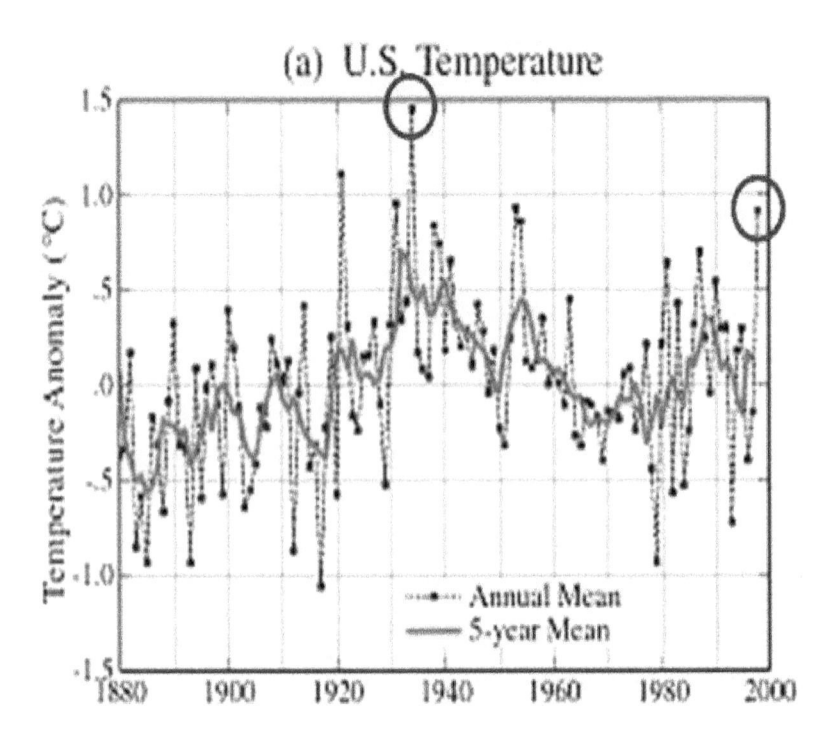

Figure 7.11
U.S. Temperature in 1999

Recently it is as follows in Figure 7-12.[231]

---
[231] Gisstemp, 2014, http://data.giss.nasa.gov/gistemp/graphs_v3/

U.S. Temperature

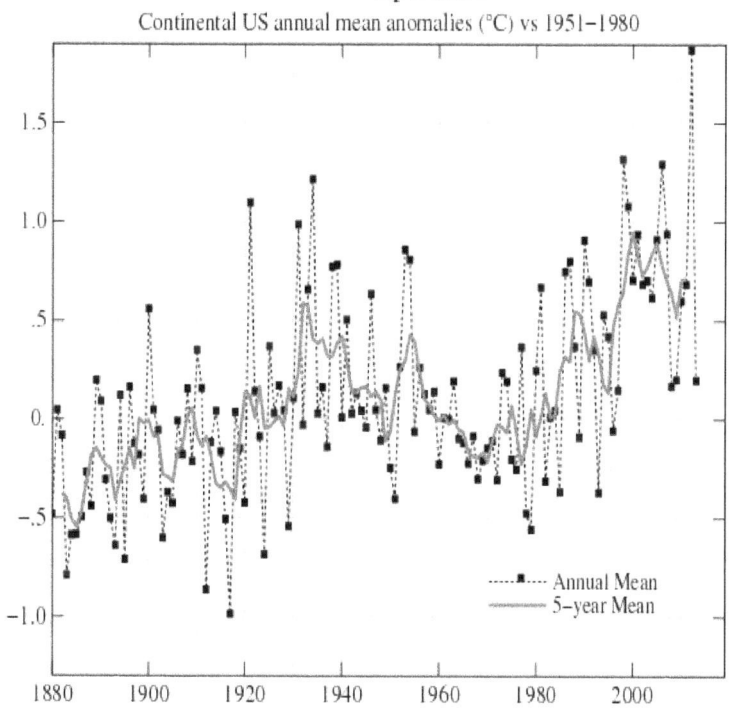

Continental US annual mean anomalies (°C) vs 1951–1980

**Figure 7.12
U.S. Temperature in 2014**

Quality of U.S. Stations

Using volunteers, Watts assessed the quality of U.S. weather stations classified by means of a quality assessment system (CRN) developed by the US NOAA.[232] His findings were as shown in Figure 7.13.[233]

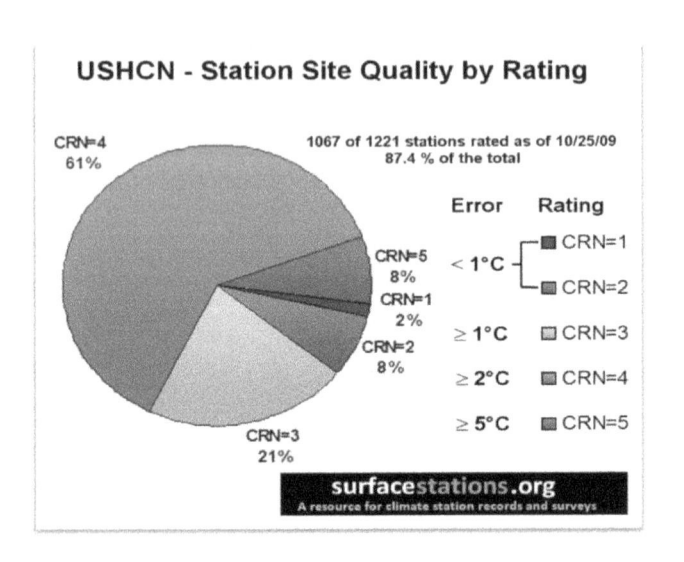

**Figure 7.13
Assessment of Quality of U.S. Surface Stations**

---

[232] D'Aleo, J., Watts, A., *2010 Surface Temperature Records Policy-Driven Deception* http://scienceandpublicpolicy.org/originals/policy_driven_deception.html

[233] Watts, A., 2009, Is *the US Surface Temperature Record Reliable?*, https://wattsupwiththat.files.wordpress.com/2009/05/surfacestationsreport_spring09.pdf

This work has led to a complete reassessment of United States weather stations. Only 2% reached the highest quality, 61% had an accuracy no better than 2°C and 8% were out by >5°C.

Watts provides a large number of photographs illustrating the factors influencing reliability.

The Hockey Stick

This title was applied to a graph purporting to give a temperature record of the Northern Hemisphere for the past 1000 years.[234]

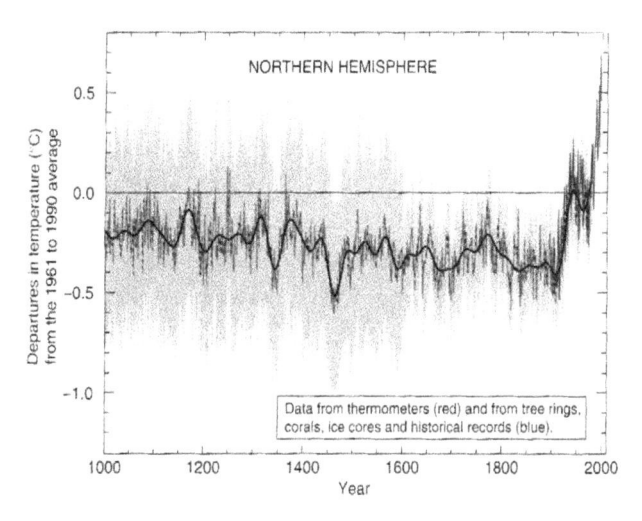

Figure 7.14
The Hockey Stick

It differed markedly from the opinions expressed in the first IPCC Report (Figure 7.2) which identified a Medieval Warm

---

[234] http://www.geo.umass.edu/faculty/bradley/mann1999.pdf

Period and a Little Ice Age.[235] This representation abolishes both of them and claims that the current MGSTAR temperature anomalies exceed temperatures for the past 1000 years by a few tenths of a degree Centigrade.

It purports to represent global temperatures for the past 1,000 years and appeared three times in the early part of the 2001 IPCC Report. *The Summary for Policymakers*, *The Technical Summary* and Chapter 2, *Observed Climate Variability and Change* in which also were several variations of it.

It depends on assuming the annual changes in various geological specimens can represent global temperature changes (proxies). Much of the proxy data is from tree ring widths. Most proxies are just for the Northern Hemisphere and on land.

It is further assumed that the principal component of variance of a time series represents temperature change.

All the assumptions are difficult to believe and involve very large uncertainties which are impossible to estimate.

The comparison with the MAGSTAR is tendentious. If there is anything in such a comparison it is most likely to involve human urban effects (which they try to deny). An assumption that it is evidence for heating effects of greenhouse gases is simply absurd.

Soon and Baliunas gathered together many of the proxies and listed them.[236] They concluded, firstly, that the coverage of data, even for the Northern Hemisphere, was not sufficiently representative to justify the deriving of an average which could be considered as reliable.[237]

---

[235] Houghton, J. T., Ding, Y., Griggs, D.J., Noguer, M., Van der Linden, P.J., Dai, X., Maskell,K., and Johnson, C.A., editors, 2001, *Climate Change 2001: The Scientific Basis*, Cambridge University Press
[236] Soon, W. and Baliunas, S., Clim. Res. 2003, 23, 89-110.
[237] Soon, W., 2005, Geophys. Res. Lett. 32 .L16712, doi:10.1029/2005GL02342.

Their second conclusion was that both the medieval warm period and the little ice age were sufficiently frequent in the observations that they must have existed. Also, there was evidence that temperatures during the medieval warm period were frequently higher than those found today.

The most devastating attacks on the hockey stick came from papers by McIntyre and McKitrick.[238] They set out to see whether they could recalculate the Mann/Bradley data and were initially surprised to find that the data were not available and had not even been supplied to the journals publishing the work.[239] The papers had been published, and believed, without any check on their validity.[240]

After a long period of wrangling they managed to get hold of most of the original data. When they carried out the calculations however, they found serious errors which, when corrected, changed the conclusion that had been attributed to them. They found that they got a higher temperature in the year 1400 than is claimed for today. They found that the shape of the curve had been automatically predetermined. The small number of actual data before 1550 led to the excessive use, including extrapolation, of several measurements which are not considered reliable by others.

Holland has documented the determined resistance of the

---

[238] Stephen McIntyre and Ross McKitrick, 2011, *Discussion of: A Statistical Analysis of Multiple Temperature Proxies: Are Reconstructions of Surface Temperatures Over the Last 1000 Years Reliable?*, Annals of Applied Statistics, Vol. 5, No. 1, 56-6 DOI: 10.1214/10-39 AOAS398L

[239] McIntyre, S. and McKitrick, R., 2003, *Corrections to Mann et al, (1998) Proxy Data Base and Northern Hemisphere Average Temperature Series*, Energy and Environment 14 751-777

[240] McIntyre, S., and McKitrick, R, 2005, *Hockey Sticks, Principal Components and Spurious Significance*, Geophys. Res. Lett. 32, L03710, doi:10.1029/2004GL02175039 39

IPCC to its acceptance of these facts.[241]

Loehle et al questioned the reliability of tree-ring measurements—which apply only to summer and are influenced by precipitation.[242] Increased temperature lowers soil moisture and the rings get thinner rather than thicker. When he used all the proxies except tree rings he got a modified record (Figure 7.14) which restored both the medieval warm period, the little ice age, and the lack of "unprecedented" character of recent temperatures.[243]

---

[241] Holland, D, 2007, *Bias and Concealment in the IPCC Process, the "Hockey Stick" Affair and its Implications*, Energy and Environment, 18 951-983.

[242] Loehle. C., 2007, *A Global Temperature Reconstruction Based on Non Tree Ring Proxies*, Energy and Environment, 18, 1049-1057.

[243] Loehle, C. and McCulloch, J. H., 2008, *Correction to a 2000-year Global Temperature Reconstruction Based on Non-tree Ring Proxies*, Energy and Environment 19, 91-100, https://notalotofpeopleknowthat.wordpress.com/2015/02/04/temperature-adjustments-transform-arctic-climate-history/

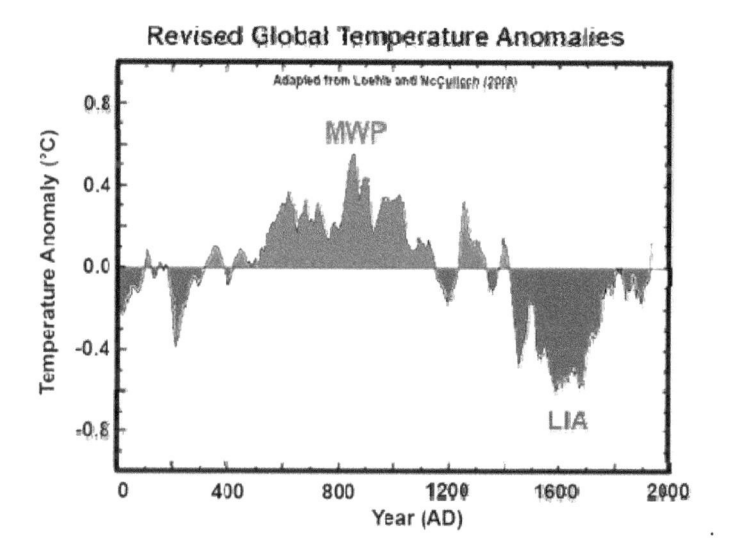

**Figure 7.15
Proxy Temperatures for the Past 1000 Years—
Ignoring Tree Rings**

Despite all of these arguments, as shown in Figure 7.15, the latest IPCC Report retains this version of the Hockey Stick. Error estimates must surely be much greater for the paleo estimates, but if the graph is taken seriously we are again involved with a few tenths of a degree of recent warming most likely caused by human urban and land use activity.

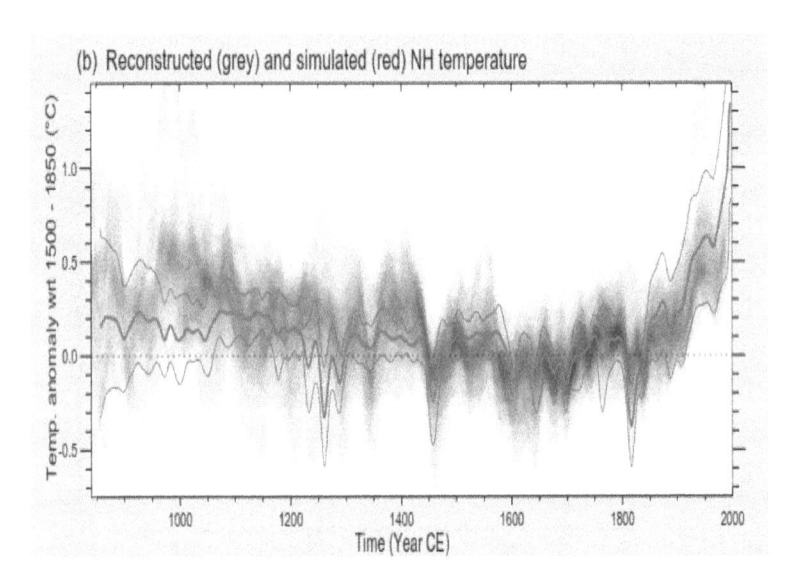

**Figure 7.15
Current IPCC Version of Reconstructed
Northern Hemisphere Temperature**

The Latest IPCC Report

These quotations from the Summary for Policymakers and Frequently Asked Questions embody most of the conclusion of the Report:

> Warming of the climate system is unequivocal, and since the 1950s, many of the observed changes are unprecedented over decades to millennia. The atmosphere and ocean have warmed, the amounts of snow and ice have diminished, sea level has risen, and the concentrations of greenhouse gases have

increased.

FAQ 2.1 | How Do We Know the World Has Warmed?

Evidence for a warming world comes from multiple independent climate indicators, from high up in the atmosphere to the depths of the oceans. They include changes in surface, atmospheric and oceanic temperatures; glaciers; snow cover; sea ice; sea level and atmospheric water vapour. Scientists from all over the world have independently verified this evidence many times. That the world has warmed since the 19th century is unequivocal.

Let us emphasize what it does not say…

Nowhere in this Report or in any previous Report is a claim that they have established a relationship between emissions of so-called greenhouse gases and the earth's climate. Nor have they shown that greenhouse gases are responsible for whatever warming is claimed.

Their claim of warming since millennia is designed to cover up the fact that for the past 18 years there has been no warming at all when estimated by their Mean Global Temperature Anomaly Record, which was the basis of their claim for warming in all the previous IPCC Reports.

They try to cover up this failure by converting it to a decadal series and by trying to argue that since global temperatures seem less important everything else is warming.

The arguments get thinner when they are scrutinised. There are no temperature measurements on ice and the thickness of ice in the Arctic Ocean is controlled by ocean oscillations beneath it.

Glaciers have many alternative influences. The total ice in the Antarctic is at a record level. The Antarctic Peninsula has an underwater volcano.

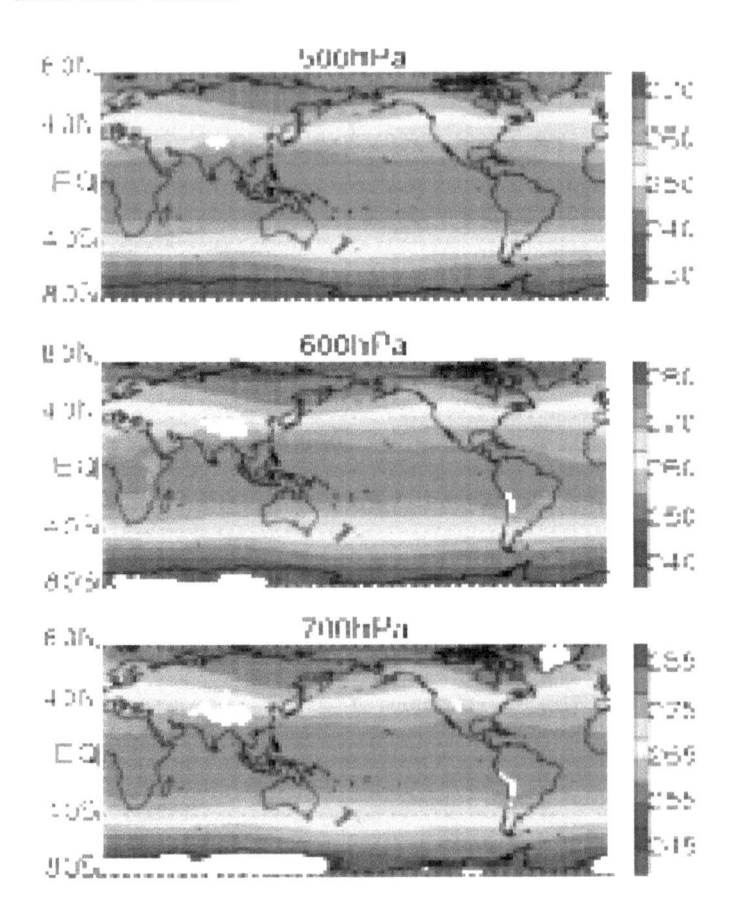

**Figure 7.16
Monthly Average Temperature Measurements
at Different Levels of the Atmosphere**

AIRS Satellite Temperatures

The Atmospheric Infrared Sounder project (AIRS) generated decadal-length, global, gridded data sets of temperature and specific humidity for several standard levels in the troposphere for the obs4MIPS project.[244] The gridded data are based on the combined retrievals from AIRS, an instrument sensitive to infrared radiation emitted from the surface and atmosphere, and the Advanced Microwave Sounding Unit (AMSU), an instrument sensitive to microwave frequencies that facilitates retrievals even in somewhat cloudy conditions. Both AIRS and AMSU are onboard.

NASA's AQUA Platform

These maps are averages over an entire column of parts of the atmosphere and they are also daily or monthly averages. They are currently unable to handle much of the detail required for weather forecasting and they are still prone to inaccuracies from the difficulties of allowing for clouds and other aerosols. They have, however, added an important extra facility for studying temperature details in the atmosphere.

Conclusions

The earth does not possess a single temperature and an acceptable average is elusive. It is therefore not possible to know whether it is warming or cooling. Rough estimates based on historical anecdotes and various proxies indicate variability over

---

[244] Atmosphere Infrared Sounder Project (AIRS), https://climatedataguide.ucar.edu/climate-data/airs-and-amsu-tropospheric-air-temperature-and-specific-humidity

geological time. Claims of global warming by the IPCC have been based on fraudulent manipulation of actual measurements and selective emphasis on favourable anecdotes.

CHAPTER 8: CARBON DIOXIDE

THERE ARE TWO gases in the earth's atmosphere without which living organisms could not exist.

Oxygen is the most abundant, 21% by volume, but without carbon dioxide, which is currently only about 0.04 percent (400PPM) by volume, both the oxygen itself, and most living organisms on earth could not exist at all.

This all began when the more complex of the two living cells (called an "eukaryote") evolved a process called a *chloroplast* some 3 billion years ago which could utilize a chemical called *chlorophyll* to capture energy from the sun and convert carbon dioxide into a range of chemical compounds and structural polymers by photosynthesis. These substances provide all the food required by the organisms not endowed with a chloroplast organelle in their cells.

This process also produced all of the oxygen in the atmosphere.

The relative proportions of carbon dioxide and oxygen have varied very widely over the geological ages as shown in Figure 8.1.[245]

---

[245] Dudley, R., 1998, *Atmospheric Oxygen, Giant Palaeozoic Insects and the Evolution of Aerial Locomotor Performance*, The Journal of Experimental

Phanerozoic atmospheric oxygen and carbon dioxide levels. From Dudley (1998).

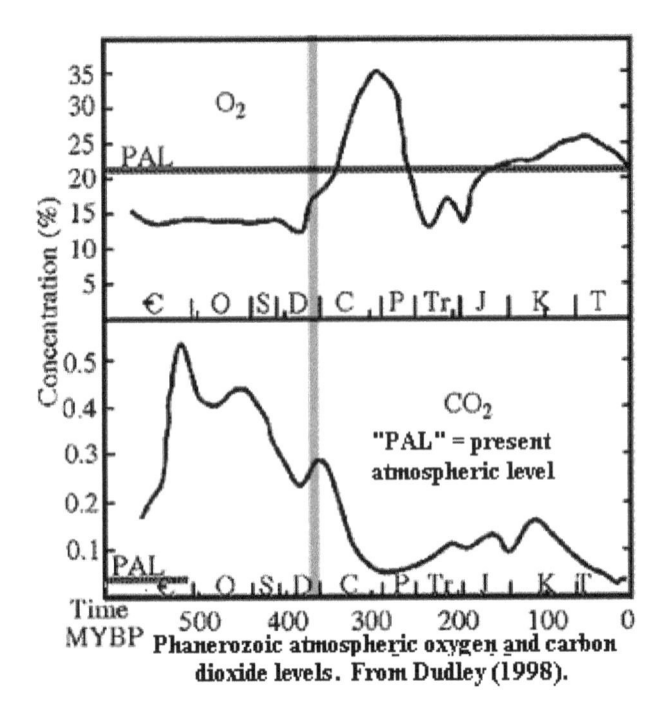

Figure 8.1
Changes in Oxygen and Carbon Dioxide Concentration in the Atmosphere over the Geological Ages

The relationship between concentration of carbon dioxide in the atmosphere and surface temperature over the geological record is shown in Figure 8.2.[246]

Biology, 201 1943-1050
http://jeb.biologists.org/content/201/8/1043.full.pdf
[246] *Plant Fossils of West Virginia*,
http://www.geocraft.com/WVFossils/Carboniferous_climate.html

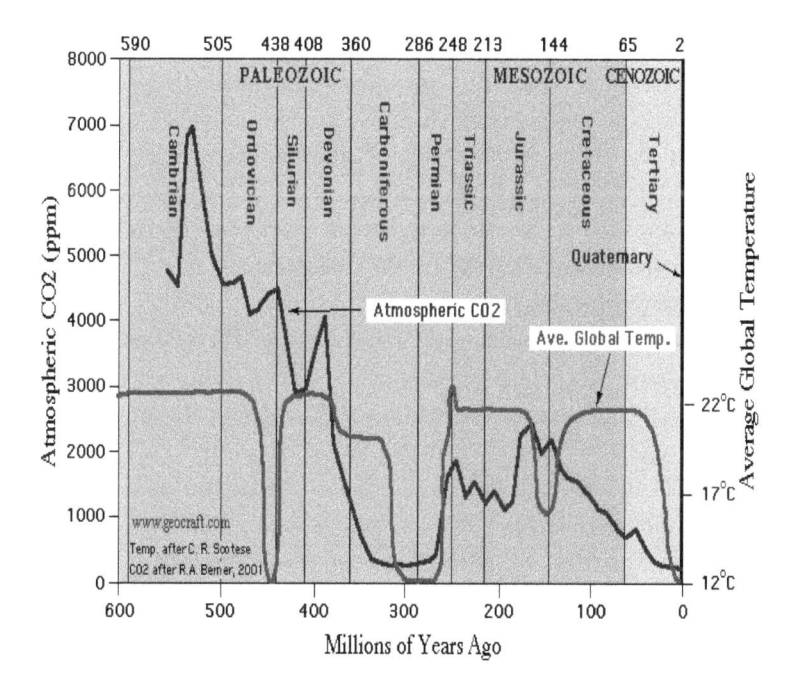

Figure 8.2
Atmospheric Concentration of Carbon Dioxide and Surface Temperature over Geological Ages

Temperature is after C.R. Scotese[247] and CO_2 after R.A. Berner and Z. Kothavala.[248]

It will be seen that there is no correlation whatsoever between atmospheric carbon dioxide concentration and the temperature at the earth's surface.

---

[247] Scotese, C. R., http://www.scotese.com/climate.htm
[248] Berner, R. A. and Kothavala, Z., 2001, *GEOCARB III: A Revised Model of Atmospheric CO_2 over Phanerozoic Time*, American Journal of Science, Vol. 301, February, 2001, pages 182-204

During the Cambrian a temperature only 10°C above today endured a maximum of 18 times the current CO_2 concentration. During most of the Palaeozoic this reduced to 10 times over today with no temperature change.

During the latter part of the Carboniferous and the Permian 250-320 million years ago, carbon dioxide concentration and temperature were similar to today but CO_2 went up to nearly 8 times what it is today when the temperature was only 10°C higher in the Jurassic.

A fall in CO_2 during the Cretaceous and Tertiary made little difference to the temperature and it is only quite recently that both have reached their present level. Oxygen in the atmosphere fluctuated from 15 to 35% during the whole period.

The theory that carbon dioxide concentration is related to the temperature of the earth's surface is therefore wrong.

The growth of plants in the Carboniferous caused a reduction in atmospheric oxygen and carbon dioxide formed the basis for large deposits of dead plants and other organisms.

Some plants of prehistory are the same ones around today. Examples are *Ginkgo biloba* and *Sciadopitys verticillata*. The temperatures must have been similar, with much greater atmospheric carbon dioxide.

Plant debris became the basis for peat and coal, while smaller organisms provided oil and gas, both after millions of years of applied heat and pressure from geological change; mountain building, erosion, deposition of sediments, volcanic eruptions, rises and fall of sea level and movement of continents. Marine organisms used carbon dioxide to build shells and coral polyps and these became the basis of limestone rocks.

Increase in atmospheric carbon dioxide caused by return to the atmosphere of some of the gas that was once there promotes the growth of forests, the yield of agricultural crops and the fish, molluscs and coral polyps in the ocean.

An increase in atmospheric carbon dioxide in the

atmosphere today is beneficial to plant life (Figure 8.3).[249]

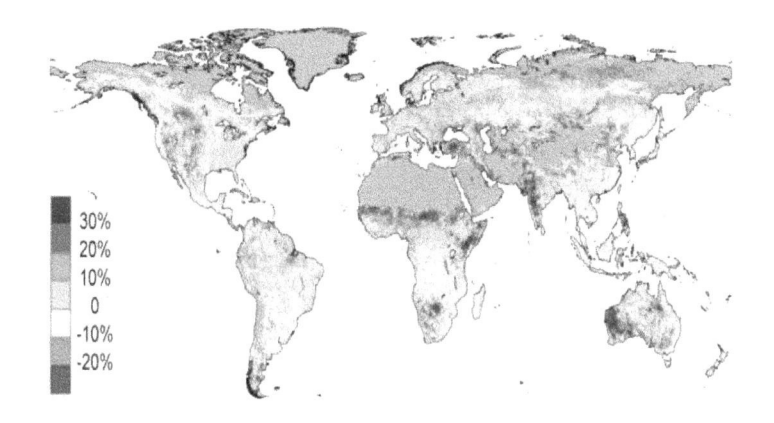

**Figure 8.3
Enhancement of Photosynthesis by Increases of Carbon Dioxide**

It is worth quoting the abstract of the paper by Randall et al 2013:

> Satellite observations reveal a greening of the globe over recent decades. The role in this greening of the "CO_2 fertilization" effect—the enhancement of photosynthesis due to rising CO_2 levels—is yet to be established. The direct CO_2 effect on vegetation should be most clearly expressed in warm, arid environments where

---

[249] Randall J. Donohue, Michael L. Roderick, Tim R. McVicar, Graham D. Farquhar, *Impact of CO_2 Fertilization on Maximum Foliage Cover across the Globe's Warm, Arid Environments*, Geophysical Research Letters, 2013; DOI: 10.1002/grl.50563

water is the dominant limit to vegetation growth. Using gas exchange theory, we predict that the 14% increase in atmospheric CO_2 (1982–2010) led to a 5 to 10% increase in green foliage cover in warm, arid environments. Satellite observations, analyzed to remove the effect of variations in precipitation, show that cover across these environments has increased by 11%. Our results confirm that the anticipated CO_2 fertilization effect is occurring alongside ongoing anthropogenic perturbations to the carbon cycle and that the fertilization effect is now a significant land surface process.[250]

Atmospheric Carbon Dioxide Measurement

The above estimates of atmospheric carbon dioxide concentration were made from a range of proxy estimates. The current claim that carbon dioxide in the atmosphere has a controlling influence on the earth's climate means that the claim depends on the accuracy with which this concentration can be measured, both in the past and in the present.

Early chemical measurements of the concentration of carbon dioxide in the earth's atmosphere have been suppressed by the Intergovernmental Panel on Climate Change.

Chapter 1 of the IPCC Fourth Report entitled *Historical overview of Climate Change Science* makes no mention of any early measurements.[251] Weart in his *History of the Carbon Dioxide*

---

[250] Ibid. *Impact of CO_2 Fertilization...*
[251] Le Treut, H., Somerville, R., Cubasch, U/, Ding, Y., Mauritzen, C., Mokssit, A., Peterson, T. and Prather, M., 2007, *Historical Overview of Climate Change, in Climate Change 2007*: The Physical Science Basis. *Contribution of Working Group I to the Fourth Assessment Report of the*

Greenhouse Effect also makes no mention of them.[252]

Beck provided an annotated list with links to internet access of almost 200 references to peer-reviewed academic scientific journal articles containing some 40,000 measurements of atmospheric carbon dioxide by chemical methods between 1800 and 1960.[253] Comprehensive data sets in more than 390 papers were ignored despite contributions from prominent scientists like Robert Bunsen, Konrad Roentgen and J. S. Haldane or the Nobel Prize winners August Krogh and Otto Warburg.

Beck lists publications of measurements of atmospheric carbon dioxide in 1816[254] and in 1830[255] by Theodore de Saussure. He was the son of Horace-Benedict de Saussure, who invented the Hot Box (which resembled a greenhouse) which was the basis of the theory of the climate developed by Jean Baptiste Joseph Fourier in 1822 and 1824 which is claimed to have originated the greenhouse effect (see Chapter 5). Yet the measurements of atmospheric carbon dioxide by de Saussure's son are completely ignored. Other early references by Letts and Blake from The Royal Dublin Society give an additional list of

---

Intergovernmental Panel on Climate Change [Solomon, S., D. Qin, M. Manning, Z. Chen, M. Marquis, K.B. Averyt, M. Tignor and H.L. Miller, editors]. Cambridge University Press, Cambridge, United Kingdom and New York, NY, USA

[252] Weart, S., 2011, *The Carbon Dioxide Greenhouse Effect*, http://www.aip.org/history/climate/co2.htm

[253] Beck, E-G, *CO2 1800-1960 Historical References*, Chemical Method, http://www.biomind.de/realCO2/literature/CO2literature1800-1960.pdf

[254] Theodore de Saussure, *Notes sur les Variations du Gaz Acide Carbonique dans L'atmosphere*, en hiver et en ete, Annales de Chimie et Physique 1816, p. 199

[255] Theodore de Saussure, 1830; *Sur les Variation de L'acide de Carbonique Atmosphérique*, Annales de Chimie et Physique, 44, p. 55

early measurements.[256]

Beck published several summaries and commentaries on the early measurements.[257] [258] [259] He also argued with Ralph Keeling.[260]

Engelbeen provided a useful summary of Beck's results where he disagrees with some of his conclusions.[261] Maijer has also been critical.[262]

Most of the early measurements were from Northern Europe. Beck considered that the earliest measurements were subject to various errors but the widespread use of more reliable equipment, particularly the Pettenkoffer titrimetric method in 1812, led to high accuracy, with a maximum 3% error reducing

---

[256] Letts, E. and Blake, R., 1899, *The Carbonic Anhydride of the Atmosphere*, Roy. Dublin Soc. Sc.Proc., N. S., Vol.9, 1899-1902; Scientific Proceedings of the Royal Dublin Society, p. 167

[257] Beck, E-G, 2007, *180 Years of Atmospheric Gas Analysis by Chemical Methods*, Energy and Environment *18 259-281*

[258] Beck E-G, *Evidence of Variability of Atmospheric CO2 Concentration during the 20th Century*, http://www.biomind.de/realCO2/literature/evidence-var-corrRSCb.pdf

[259] Beck E-G, Comment + reply from author to R F Keelng and H Mrijer on "180 Years of atmospheric CO_2 gas analysis by chemical methods by" by Ernst-Georg Beck, Energy and Environment, *Vol. 18(2), 259-282, 2007* http://www.biomind.de/treibhaus/180CO2/author_reply9-2.pdf

[260] Beck E-G Reply to "Comments on 180 Years of Atmospheric CO_2 Gas Analysis by Chemical Methods." Energy & Environment, Vol. 18(2), 2007

[261] Engelbeen, F., *2010 Historical CO2 Measurements by Ernst Beck* http://www.ferdinand-engelbeen.be/klimaat/beck_data.html

[262] Meijer, H., www.biomind.de/.../180CO2/Comment_E&E-on_Beck_Meijer_update.d..http://www.biomind.de/treibhaus/180 CO2/Comment_E&E-on_Beck_Meijer_update.doc

to 1% for the data of Henrik Lundegardh.[263][264]

The measurements selected by Beck were from rural areas or the periphery of towns, under comparable conditions of a height of approximately 2 meters above ground at a site distant from potential industrial contamination. They showed a variation with time of day, of season, and of wind speed and direction, making it difficult to derive a local average. There were frequent measurements of concentrations higher than those reported as background concentrations by NOAA at present.

These measurements were made by real people with proper instruments in a large number of localities. They give a much better appreciation of variability and change in atmospheric carbon dioxide concentration over the period than the deductions from gas trapped in ice cores which are from unrepresentative locations and subject to much uncertainty.[265]

In 1958, Charles Keeling introduced a new technique for the accurate measurement of atmospheric CO_2 using cryogenic condensation of air samples followed by NDIR spectroscopic analysis against a reference gas, using manometric calibration.

Subsequently, this technique was adopted as an analytical standard for CO_2 determination throughout the world, including by the World Meteorological Association.

The climate models sponsored by the Intergovernmental

---

[263] Lundegardh, H., *Der Kreislauf der Kohlensäure in der Natur*, Fischer, Jena (680), 1924,
http://www.biokurs.de/treibhaus/literatur/Lundegardh/lundegardh2.doc

[264] Lundegardh, H., *Klima und Boden und ihre Wirkung auf das Pflanzenleben*, Jena, 1949,
http://www.archive.org/stream/zeitschriftfrb16jena#page/n271/mode/2up

[265] Jaworowski, Z., 2007, *CO2: The Greatest Scientific Swindle of Our Time*, EIR Science (March), 38-55

Panel on Climate Change are based on the belief that the global climate has a balanced energy which is only changed by increasing concentrations of carbon dioxide and other greenhouse gases. These gases are assumed to be well-mixed so that their concentration, all over the world, is a constant at any one particular time, only increasing with human emissions.

In order to support this theory, Keeling at the Scripps Institution of Oceanography, discovered that there was an almost consistent background concentration of carbon dioxide which could be considered to apply globally and identified from suitable sites which could be shown to increase with carbon dioxide emissions.

The procedure required to indentify this *background* is described in some detail by Tans and Thoning for the observatory at Mauna Loa.[266]

Measurements whose standard deviation fell below a specified minimum were rejected. On average, over the entire record, there are 13.6 retained hours per day with background CO_2. The rest were rejected as noise.

Beck has discussed the Mauna Loa measurements.[267] Comparison between old wet chemical and new physical methods in 1958 and 1967 on sea and land give a difference of about +10PPM for the new procedure.

A similar procedure has been described for New Zealand.[268]

At Baring Head maritime well mixed air masses come from the Southerly direction, and a baseline event is normally defined

---

[266] Pieter Tans and Kirk Thoning, *How We Measure Background CO2 at Mauna Loa*,
http://www.esrl.noaa.gov/gmd/ccgg/about/co2_measurements.pdf
[267] Beck E-G, 2008, *50 Years of Continuous Measurement of CO2 on Mauna Loa*, Energy and Environment 19 No. 7
[268] Manning, M. R., Gomez, A.J and Pohl, K.P., *Trends*,
http://cdiac.ornl.gov/trends/co2/baring.html

as one in which the local wind direction is from the South and the standard deviation of minute-by-minute CO_2 concentrations is <<0.1PPMV for 6 or more hours.

This background concentration is supposed to be well-mixed and to be unaffected by sources and sinks.

Yet the oceans are themselves contaminated with sources and sinks (see Figure 8.4).[269]

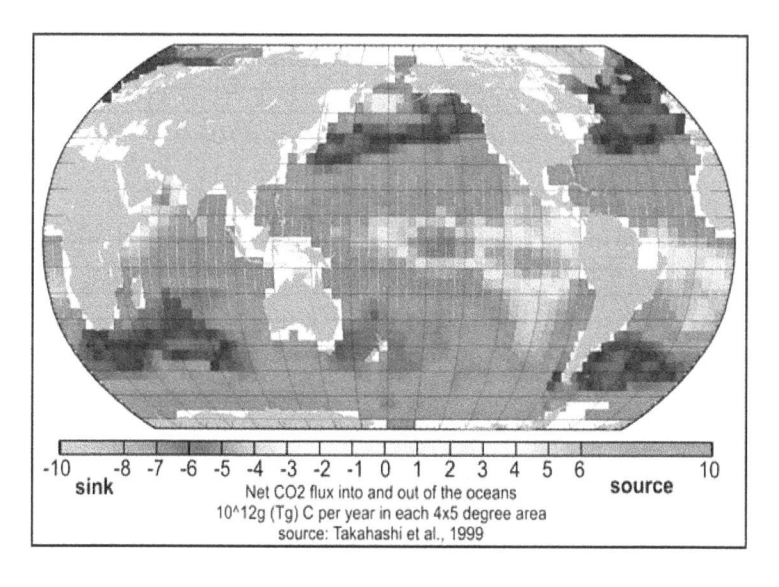

Figure 8.4
CO_2 Flux for the Oceans

---

[269] Takahashi T et al.,1999, *Deep-Sea Research II 49 (2002) 1601-1622* Global sea-air CO2 Flux based on Climatological Surface Ocean CO2, and seasonal biological and temperature effects
http://www.ldeo.columbia.edu/~csweeney/papers/taka2002.pdfhttp://www.seafriends.org.nz/issues/global/acid2.htm

Vincent Gray

The region around Mauna Loa includes areas with CO_2 emissions, and much of the rest is a sink. It is understandable how difficult it is to get a sufficiently constant sample.

In order to claim that there is such a thing as a background CO_2 it has been necessary to ensure that all measurements everywhere in the world are made from samples from over the oceans. Measurements over land surfaces have been comprehensively discouraged.

Yet the greenhouse effect is about emissions, namely contamination. It is crazy to take all this trouble to make measurements which do not involve the emitted gases themselves, but only a small fraction that is considered to be well-mixed, then to claim that it is these background figures apply to the entire atmosphere.

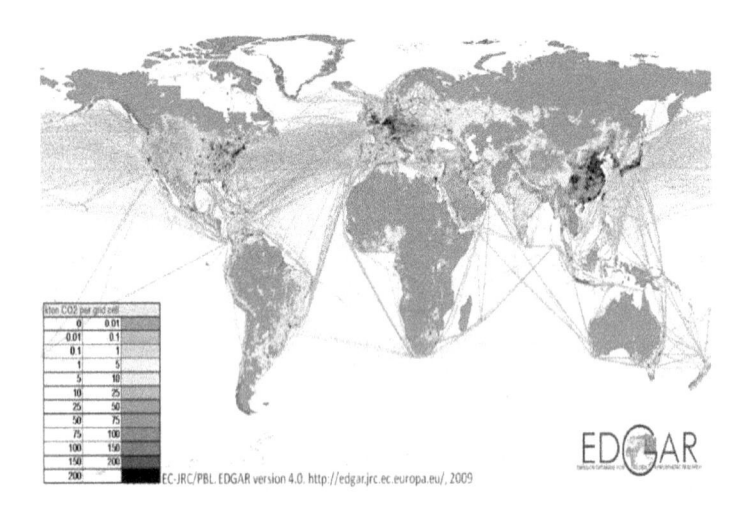

**Figure 8.5
Distribution of Carbon Dioxide Emissions**

The map shown in Figure 8.5 illustrates that actual local concentrations of carbon dioxide are greatest over the three large industrial areas.[270] Since the supposed greenhouse effect is dependent on the logarithm of the carbon dioxide concentration, increases above these areas is negligible or zero and the main supposed effects are on the areas with low current concentrations where it encourages plant growth.

This map does not tell the whole story—satellite measurements of carbon dioxide levels in the atmosphere have recently improved with the Atmospheric Infra Red Sounder (AIRS) and NASA's Aqua level 3 satellite, which is able to provide monthly figures for mid troposphere concentration.[271]

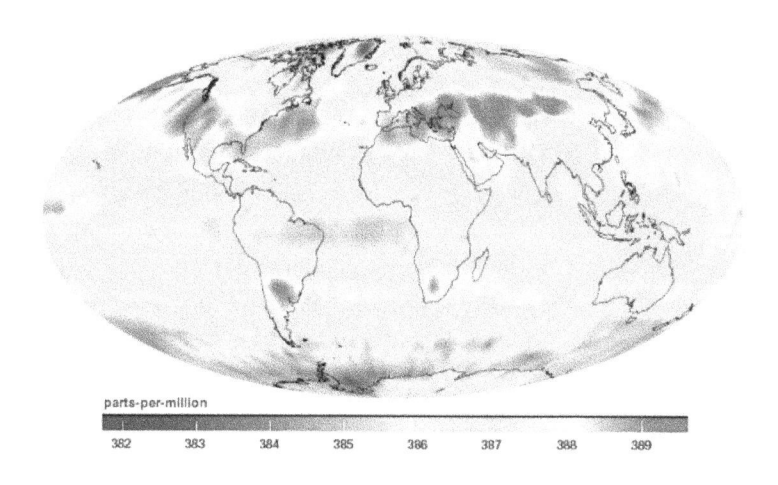

Figure 8.6
Average Carbon Dioxide Concentration in the Mid Troposphcrc July 2009

---

[270] EDGAR, *Emission Database for Global Atmospheric Research*, http://edgar.jrc.ec.europa.eu/part_CO2.php
[271] Climate Change Indicators, http://scentofpine.org/indicators/

This map shows that for the mid troposphere regions the high emissions from the industrial countries slowly circulate, so that they are no longer above the regions of emission.[272] Since this is a time as well as a column average the actual carbon dioxide concentration at any small region in the atmosphere is changing all the time and an overall figure above a particular place on the earth is continuously varying and currently unpredictable.

It also means that background measurements are no longer relevant, as are any measurements made only on the surface. If extra emissions have an effect on the climate those places where the CO_2 concentration is high will have little or no change because of the logarithmic relationship. Rural and lower CO_2 places would have the greatest changes from an increase and has been shown in Figure 8.3, this is likely to be beneficial. So, carbon dioxide is not well-mixed in the atmosphere and the overall global models need to be modified to allow for regional CO_2 change as they have already been modified to allow for regional temperature and precipitation.[273]

NASA has even provided an animated video based on a model of what they think happens.[274] It shows that actual carbon dioxide concentrations vary with time and level everywhere in the atmosphere. The new OC-2 satellite[275] promises to make individual time- and level-based measurements.

---

[272] Freeman, T., 2009, *Inside Copenhagen: The Challenge of Atmospheric Bookkeeping*, http://climate.nasa.gov/blog/226

[273] IPCC, 2013, *Annex 1: Atlas of Global and Regional Climate Projections*, van Oldenborgh, G.J., M. Collins, J. Arblaster, J.H.Christensen, J. Marotzke, S.B. Power, M. Rummukainen and T. Zhou, editors, *Climate Change 2013: The Physical Science Basis. Contribution of Working Group I to the Fifth Assessment Report of the Intergovernmental Panel on Climate Change*, Stocker,T.F.,D.Qin,G.

[274] NASA, *A Year in the Life of Earth's CO2*, https://www.youtube.com/watch?v=x1SgmFa0r04

[275] Orbiting Carbon Observatory OCO-2 http://oco.jpl.nasa.gov/

A comparison between Figures 8.5 and 8.6 shows that there is a strong correspondence between regions with high CO_2 emissions in Figure 8.5 and regions with higher CO_2 concentration in Figure 8.6s. It is therefore true that the additional CO_2 in the atmosphere is largely caused by CO_2 emissions.

There are two stable nonradioactive isotopes of carbon C_{12} about 98.9% and C_{13} about 1.1% Since there is a slight difference in the ratio between them for C_3 and C_4 plants whose difference in physiology are used in attempts to prove that decreases in the proportion of C_{13} is due to fossil fuel combustion. This is now unnecessary as the AIRS map (Figure 8.6) proves it is true.

Measurement of Downward Intensity

There are now measurements of downward radiation from regions absorbing radiation from the earth in the region associated with carbon dioxide. An example is shown in Figure 8.7.[276]

---

[276] *CO2 Forcing*,
 http://wattsupwiththat.com/2015/02/25/almost-30-years-after-hansens-1988-alarm-on-global-warming-a-claim-of-confirmation-on-co2-forcing/

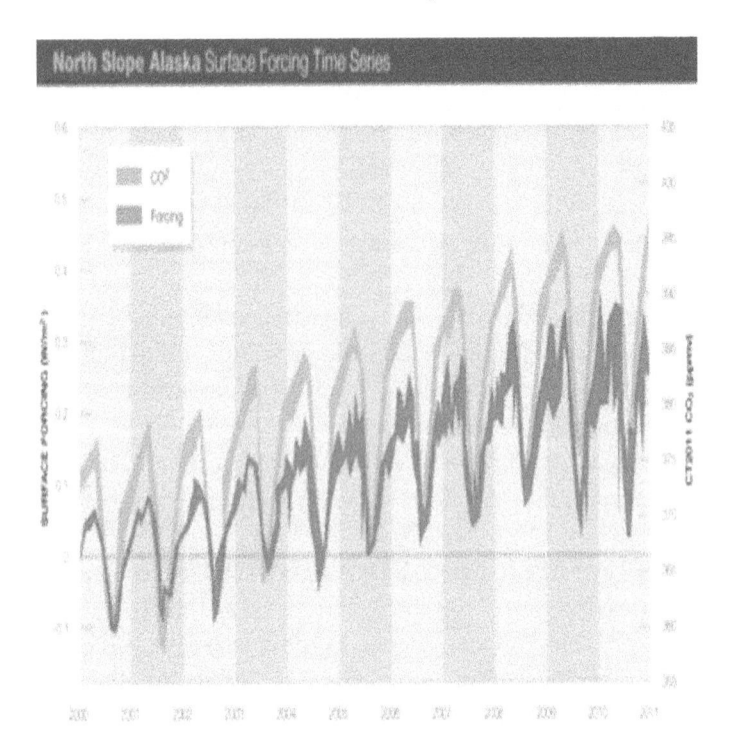

Figure 8.7
Downward Radiation Intensity from Atmospheric Carbon Dioxide

Conclusions

Estimates of carbon dioxide concentration in early atmospheres show that it has no relationship with the local surface temperature.

Recent satellite measurements of column concentrations of atmospheric carbon dioxide show that the gas is not well-mixed in the atmosphere and is not characterised by the so-called background concentration. The variability, combined with the logarithmic effect of concentration on temperature, means that additional emissions would have little or no effect over industrial areas and a largely beneficial effect elsewhere. The overall warming effect would therefore be much lower than claimed current estimates.

Vincent Gray

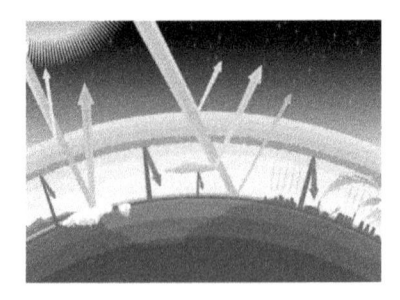

CHAPTER 9: CLIMATEGATE EMAILS

In November 2009, a large number of emails and other material from the Climate Research Unit of the University of East Anglia in the UK suddenly became available on the Internet.[277] [278] [279] The mysterious hacker or whistleblower has never been revealed—so far.

They amounted to a total of 120MB, 1078 emails from 6/03/1996 to 13/11/2009 and 72 documents, computer code and models. They dealt with the exchanges and activities of several of the most senior scientists engaged in the climate change swindle over this period.

Every effort has been made to suppress them. Several

---

[277] *Climategate emails*,
http://assassinationscience.com/climategate/1/FOIA/mail/
[278] Tom Nelson,
http://tomnelson.blogspot.co.nz/p/climategate_05.html
[279] *Climategate*
http://www.telusplanet.net/dgarneau/climate-e-mails.htm
http://metis-history.info/climate-e-mails.shtml
http://junksciencearchive.com/FOIA/

servers I have tried ignore requests to supply the actual emails and instead gave many pages of articles devoted to attempts to whitewash them or explain them away. Most of the earlier websites that supplied them have either disappeared or been hacked.

At the time of writing this they could be accessed by copying the references into your server. A good summary of the most important ones is supplied by the Canadians.[280] A set of links to the most important documents and comments has been made by Anthony Watts.[281] Further useful summaries are by Costella[282][283] and Monckton.[284]

The emails provide detailed information about how this group of scientists were able to manipulate the imposition on the world public of their false theory of the climate. It involved distorting and fabricating climate data, intimidating opposition using every technique of public relations, dishonest advertising techniques and spin and efforts to prevent criticism from being published or debated. They showed how they exercise control

---

[280] Ibid. Climategate
[281] Watts Up With That,
 http://wattsupwiththat.com/climategate/
http://wattsupwiththat.com/2012/01/06/250-plus-noteworthy-climategate-2-0-emails/
[282] Costella, J. P., *Climategate: A Step by Step Analysis*
http://heartland.org/policy-documents/climategate-step-step-analysis
[283] Costella, J. P ,
http://www.lavoisier.com.au/articles/greenhouse-science/climate-change/climategate-emails.pdf
http://scienceandpublicpolicy.org/images/stories/papers/reprint/climategate_analysis.pdf
[284] Monckton, *Climategate Caught Green-Handed*,
http://scienceandpublicpolicy.org/originals/climategate.htm

over journal editors and of the peer review process and obstruction from official information requests.

After conversation at the popular Anthony Watts' WUWT website, Telegraph journalist James Delingpole wrote a blog called *Climategate: the Final Nail in the Coffin of 'Anthropogenic Global Warming'?* and the Climategate meme went viral.[285] The coining of the name "Climategate" and the evolution of the meme was researched by student David Norton.[286] Christopher Booker called it: *The Worst Scandal of our Generation.*[287]

From this enormous store I can only quote a few. Let us start with extracts from Delingpole's article:

Manipulation of Evidence

> *I've just completed Mike's Nature trick of adding in the real temps to each series for the last 20 years (ie from 1981 onwards) amd from 1961 for Keith's to hide the decline. (This is further treated by Just Fact)*

Fantasies of violence against prominent Climate Sceptic scientists:

> *Next time I see Pat Michaels at a scientific*

---

[285] James Delingpole, http://blogs.telegraph.co.uk/news/jamesdelingpole/100017393/climategate-the-final-nail-in-the-coffin-of-anthropogenic-global-warming/
[286] http://bigthink.com/age-of-engagement/student-researcher-tracks-the-origins-of-the-climategate-name
[287] Booker Christopher, http://www.telegraph.co.uk/comment/columnists/christopherbooker/6679082/Climate-change-this-is-the-worst-scientific-scandal-of-our-generation.html

> meeting, I'll be tempted to beat the crap out of him. Very tempted.

Attempts to disguise the inconvenient truth of the Medieval Warm Period (MWP):

> Phil and I have recently submitted a paper using about a dozen NH records that fit this category, and many of which are available nearly 2K back—I think that trying to adopt a timeframe of 2K, rather than the usual 1K, addresses a good earlier point that Peck made w/ regard to the memo, that it would be nice to try to "contain" the putative "MWP", even if we don't yet have a hemispheric mean reconstruction available that far back....

And, perhaps most reprehensibly, a long series of communications discussing how best to squeeze dissenting scientists out of the peer review process. How, in other words, to create a scientific climate in which anyone who disagrees with AGW can be written off as a crank whose views do not have a scrap of authority.

This was the danger of always criticizing the skeptics for not publishing in the "peer-reviewed" literature. Obviously, they found a solution to that—take over a journal! So what do we do about this?

> I think we have to stop considering "Climate Research" as a legitimate peer-reviewed journal. Perhaps we should encourage our colleagues in the climate research community to no longer submit to, or cite papers in, this journal. We

> would also need to consider what we tell or request of our more reasonable colleagues who currently sit on the editorial board...What do others think?
>
> I will be emailing the journal to tell them I'm having nothing more to do with it until they rid themselves of this troublesome editor.
>
> It results from this journal having a number of editors. The responsible one for this is a well-known skeptic in NZ. He has let a few papers through by Michaels and Gray in the past. I've had words with Hans von Storch about this, but got nowhere. Another thing to discuss in Nice.[288]

This email mentions myself. I submitted a paper entitled *The IPCC Future Projections: Are They Plausible* to the Journal *Climate Research* in October 1990 It was published in August 14th, 1991 as Vol 19 155-162, 1991.[289]

As a result of the above email, the Editor, Chris de Freitas, who had also approved a paper by Willie Soon, was sacked—and so were the entire Editorial Board. Since then, *Climate Research* will not permit any criticism of the greenhouse scam.

Kevin Trenberth said:

> Well I have my own article on where the heck is global warming? We are asking that here in Boulder where we have broken records the past two days for the coldest days on record. The fact is that we can't account for the lack of warming

---

[288] *Just Facts*, http://www.justfacts.com/globalwarming.hidethedecline.asp
[289] Gray, V.R., *The IPCC Scenarios: Are They Plausible?*, Climate Research 19 155-162, 1991

at the moment and it is a travesty that we can't. The data published in the August 2009 supplement on 2008 shows there should be even more warming: but the data are surely wrong. Our observing system is inadequate.

How come you do not agree with a statement that says we are nowhere close to knowing where energy is going or whether clouds are changing to make the planet brighter? We are not close to balancing the energy budget. The fact that we cannot account for what is happening in the climate system makes any consideration of geoengineering quite hopeless as we will never be able to tell if it is successful or not! It is a travesty!

Here are some of the issues as I see them: Saying it is natural variability is not an explanation. What are the physical processes? Where did the heat go? But the resulting evaporative cooling means the heat goes into atmosphere and should be radiated to space: so we should be able to track it with Clouds and the Earth's Radiant Energy System data. That data are unfortunately wanting and so too are the cloud data. The ocean data are also lacking although some of that may be related to the ocean current changes and burying heat at depth where it is not picked up. If it is sequestered at depth then it comes back to haunt us later and so we should know about it.[290]

October 14: Tom Wigley replies

> Kevin, I didnt mean to offend you. But what you said was "we cant account for the lack of warming at the moment". Now you say "we are nowhere close to knowing where energy is going". In my eyes these are two different things—the second relates to our level of understanding, and I agree that this is still lacking.
>
> We are now debating how quickly the ship is sinking. But why didnt any of these scientists speak up when their paymasters said to the world, "the science is settled"?

So Tom Wigley believes that the *ship is sinking*. Let us take a look at his other views.

August 1999. Tom Wigley, former director of (CRU) Climate Research Unit (1978-1993), who starting in 2009 works for the (NCAR) National Center for Atmospheric Research in Boulder, Colorado, believes ice cores were unreliable because they correlate very poorly with (land) temperature. He said the link between ice core and temperature variation was "close to zero" and tree rings were less than 50% reliable (to other evidences?). The main external candidate is solar, and more work is required to improve the 'paleo' solar forcing record. The OIS-3 studies using ice cores suggested climate and temperatures on continental lands are poorly known, due to the discontinuous nature of sedimentation changes on land. This

---

[290] *Climate Sanity*, https://climatesanity.wordpress.com/2009/11/24/kevin-trenberths-real-travesty/

The Global Warming Scam

scientific nonsense must stop if a political objective is to be met.

> *1997 November 25: From: Tom Wigley To: jan.goudriaan, grassl_h, Klaus Hasselmann, Jill Jaeger, oriordan, uctpa84, john, mparry, pier.vellinga*
> *Subject: Re: ATTENTION. Invitation to influence Kyoto.*
>
> *I was very disturbed by your recent letter, and your attempt to get others to endorse it. Not only do I disagree with the content of this letter, but I also believe that you have severely distorted the IPCC "view" when you say that "the latest IPCC assessment makes a convincing economic case for immediate control of emissions." In contrast to the one-sided opinion expressed in your letter, IPCC WGIII SAR and TP3 review the literature and the issues in a balanced way presenting arguments in support of both "immediate control" and the spectrum of more cost-effective options. It is not IPCC's role to make "convincing cases" for any particular policy option; nor does it. However, most IPCC readers would draw the conclusion that the balance of economic evidence favors the emissions trajectories given in the WRE paper. This is contrary to your statement. This is a complex issue, and your misrepresentation of it does you a dis-service. To someone like me, who knows the science, it is apparent that you are presenting a personal view, not an informed, balanced scientific assessment. What is unfortunate is that this will*

not be apparent to the vast majority of scientists you have contacted. In issues like this, scientists have an added responsibility to keep their personal views separate from the science, and to make it clear to others when they diverge from the objectivity they (hopefully) adhere to in their scientific research. I think you have failed to do this.

Your approach of trying to gain scientific credibility for your personal views by asking people to endorse your letter is reprehensible. No scientist who wishes to maintain respect in the community should ever endorse any statement unless they have examined the issue fully themselves. You are asking people to prostitute themselves by doing just this! I fear that some will endorse your letter, in the mistaken belief that you are making a balanced and knowledgeable assessment of the science—when, in fact, you are presenting a flawed view that neither accords with IPCC nor with the bulk of the scientific and economic literature on the subject.

When scientists color the science with their own PERSONAL views or make categorical statements without presenting the evidence for such statements, they have a clear responsibility to state that that is what they are doing. You have failed to do so. Indeed, what you are doing is, in my view, a form of dishonesty more subtle but no less egregious than the statements made by the greenhouse skeptics, Michaels, Singer et al. I find this extremely disturbing.

1999 May 19: Tom Wigley writes to Mike Hulme and Mike MacCracken, regarding a chain of emails discussing climate models:

> I've just read the emails of May 14 onwards regarding carbon dioxide. I must say that I am stunned by the confusion that surrounds this issue. Basically, I and MacCracken are right and Felzer, Schimel and (to a lesser extent) Hulme are wrong. There is absolutely, categorically no doubt about this.

Mike Hulme responds:

> I still have a problem making sense of what the Met(eorological) Office Hadley Centre have published.

Tom Wigley replies:

> Yes, I am aware of the confusion surrounding what the Met(eorological) Office Hadley Centre did and why. It is even messier than you realize. The Hadley people have clearly screwed things up, but their "errors" dont really matter given all of the uncertainties. I didnt mention this because I thought that opening up that can of worms would confuse people even more. The climate model output is also uncertain.

2009 October 5: Tom Wigley wrote to Phil Jones, CRU head, to say that sceptic Steve McIntyre was actually right, CRU deputy director Keith Briffa had made an extraordinary "mess" of tree ring data which he'd claimed showed the world hadn't been hotter. Wigley also wondered why Briffa had chosen just 12 trees in Yamal to show modern warming, and failed to include a much larger sample which would have shown cooling instead. He also warned against the CRU's hiding of data:

On October 5, Tom Wigley wrote to Phil Jones:

> *It is distressing to read that American Stinker item. But Keith Briffa does seem to have got himself into a mess. As I pointed out in emails, Yamal (Siberia) is insignificant. I presume they went thru papers to see if Yamal was cited, a pretty foolproof method if you ask me. Perhaps these things can be explained clearly and concisely—but I am not sure Keith is able to do this as he is too close to the issue and probably quite pissed of [sic]. I think Keith needs to be very, very careful in how he handles this. Id be willing to check over anything he puts together.*
>
> *Phil Jones forwarded the email to Keith*

There is more. I have already quoted the one where Wigley agrees that Wei-Chyung Wang is a fraud (Chapter 7).

Phil Jones was Director throughout the period of the emails. Wigley, the previous Director, was the *eminence grise* of the first two IPCC Reports (if you include the original Final Draft of the second) when the view expressed officially was that the warming could be natural.

Although he is shown to be keen on preventing publication from sceptics, Wigley frequently expresses disagreement with much of what is being done by his former staff and their friends since he left. He is surely the Prime Suspect behind the release of these emails.

Kevin Trenberth took over the leading theoretical role for

the subsequent three IPCC Reports, but he also is obviously uncomfortable with what he feels impelled to say.

Then Mike Mann explains what it's all about:

email Sept 1992:

> *I am definitely using the version of the Briffa et al series you sent in which Keith had restandardized to retain more low-frequency variability relative to the one shown by Briffa et al (1998). So already, the reconstruction I'm using is one-step removed from the published series (as far as I know!) and that makes our use of even this series a bit tenuous in my mind, but I'm happy to do it and let the reviewers tell us if they see any problem.*

Mike 2009, October 27, Mike Mann to Phil Jones and Gavin Schmidt:

> *As we all know, this isn't about truth at all, its about plausibly deniable accusations.*

And again:

> *...it's tough when even your allies are starting to turn:*
> > *Be a bit careful about what information you send to Andy Revkin [of the New York Times] and what emails you copy him in on. He's not as predictable as we'd like...*

An interesting exchange is a correspondence between Keith Briffa whose results differed from the views of Eugene Wahl.[291]

So, Briffa writes confidentially to Wahl for help and Wahl assists him by passing a copy of a paper that has yet to be published. The aim is to answer concerns that McIntyre as reviewer has raised. Wahl and Amman's words are incorporated in the response to McIntyre with the hope that no one will ever notice.

Two years later, someone does notice. It's May 24th 2008, and Steve McIntyre, climate science puzzle solver, is reading the reviewer comments to chapter 6 of AR4 written in 2006. In the course of reviewing Briffa's replies to him, McIntyre notes something peculiar. Briffa's replies, written in 2006, seemed to plagiarize an unpublished paper by Casper Amman and Eugene Wahl published in 2007. That is, in 2006, Briffa was repeating the argument of a paper that was not published until 2007. How could Briffa plagiarize an article that hadn't been published? Why would he repeat the arguments almost word for word? Who was feeding Briffa his arguments? How was Briffa doing this if all communication with the authors had to be part of the official record?

At the time, in May of 2008, McIntyre assumed that Briffa was getting information from Casper Ammann since Ammann was listed as a contributing author to chapter 6. It did not occur to McIntyre that Wahl was the source of the text. Thanks to the individual who liberated the Climategate emails, we now know that Wahl was the source of that text. The Climategate emails, quoted above, show Briffa and Wahl exchanging emails about the way McIntyre's arguments should be handled. Confidentially, outside the process of the IPCC which is designed to capture

---

[291] Briffa/Wahl correspondence,
http://www.prisonplanet.com/climategate-mann-ordered-wahl-to-delete-emails.html

reviewer objections and authors' responses to those objections. Wahl is brought in by Briffa to defend his own work and defend it with literature that has not been published yet.

At the same time in 2008, across the ocean, David Holland had been reading McIntyre's work and he had issued an FOIA request to the Climatic Research Unit–CRU. That FOIA request covered all correspondence coming in and out of CRU relative to chapter 6 of AR4. The hunt for the source that was feeding Briffa was on, with Holland leading the charge. At CRU, FOIA officer Palmer instructs the team that they must do everything "by the book" because Holland will most certainly appeal a rejection letter.

In that context, Jones writes the famous email to Mann. Jones requests that Mann delete his emails and he requests that Mann contact Wahl and have Wahl delete his emails. Is Jones covering his bases in case of an appeal? Is he covering his bases against an FOIA request that might be served on Mann and Wahl in the U.S.? In any case, he appears to be conspiring with others to deny Holland his FOIA rights.

> Mike,
> Can you delete any emails you may have had with Keith re AR4? Keith will do likewise. He's not in at the moment – minor family crisis Can you also email Gene and get him to do the same? I don't have his new email address. We will be getting Caspar to do likewise.
> I see that CA claim they discovered the 1945 problem in the Nature paper.

Mann responds that he will contact Wahl ASAP, which he does.

> Hi Phil, laughable that CA would claim to have

> *discovered the problem. They would have run off to the Wall Street Journal for an exclusive were that to have been true. I'll contact Gene about this ASAP. His new email is:*
> *generwahl@xxxxxxxxx.xxx*
> *talk to you later,*
> *mike*

As Wahl told the investigators in 2011, Mann contacted him and Wahl deleted his emails.

In 2010, in an effort to clear Mann of any wrong doing, a committee of inquiry was set up at Penn State. We now know that committee failed miserably. They failed for many reasons, but the Wahl admission is the starkest example.

Here is one allegation the committee investigated:

> *Allegation 2: Did you engage in, or participate in, directly or indirectly, any actions with the intent to delete, conceal or otherwise destroy emails, information and/or data, related to AR4, as suggested by Phil Jones?*
>
> *Finding 2. After careful consideration of all the evidence and relevant materials, the inquiry committee finding is that there exists no credible evidence that Dr. Mann had ever engaged in, or participated in, directly or indirectly, any actions with intent to delete, conceal or otherwise destroy emails, information and/or data related to AR4, as suggested by Dr. Phil Jones. Dr. Mann has stated that he did not delete emails in response to Dr. Jones' request. Further, Dr. Mann produced upon request a full archive of his emails in and around the time of the preparation of AR4. The archive contained*

e-mails related to AR4.

The committee found this because they apparently failed to understand Mann's reply. As they reported:

> *He* [Mann] *explained that he never deleted emails at the behest of any other scientist, specifically including Dr. Phil Jones, and that he never withheld data with the intention of obstructing science;* ...

What can we make of this? Mann was apparently asked the question:

> *Did you engage in or participate in, directly or indirectly, any actions with the intent to delete emails?*

And it seems clear he only answered half of the question, leaving the unanswered second part dangling: did you contact anyone or otherwise 'indirectly' participate in deleting records? This either did not strike, or did not interest, the Penn State 'investigators.' This despite that Mann, it appears, answered "carefully" and incompletely. He only answered that *he* hadn't deleted emails. He never directly denies partaking, indirectly, in the deletion of Wahl's emails. He apparently withheld the information that he had asked Wahl to delete emails.

Is this a lie? Not directly. It's more what Wikipedia would describe as "Careful Speaking."

Careful Speaking is distinct from the above in that the speaker wishes to avoid imparting certain information or admitting certain facts and, additionally, does not want to 'lie' when doing so. Careful speaking involves using carefully phrased statements to give a 'half-answer': one that does not actually 'answer' the question, but still provides an appropriate (and accurate) answer based on that question. As with 'misleading,' below, 'careful speaking' is not outright lying.

So why did the inquiry, stocked with Mann's fellow professors, fail to ask good follow up questions? We really do not know because we don't have access to the transcript of their interview with Mann. Did he intend to deceive? Or did he just speak "carefully?" It would seem that the actual transcript of the questions and answers should be published. Perhaps Congress should serve the members of the inquiry with a subpoena. That would allow people to decide if Mann lied or if he just spoke carefully.

And there are a few more questions we need to ask. Mann claims that he never deleted the emails. But he asked Wahl to delete the emails. This makes no sense that Mann would participate in a cover up by passing along a message to another participant of that cover-up downstream and not delete emails himself. It defies any logical reconstruction of events. Why would Mann ask Wahl to do something that he himself would not do? We also know from the inquiry that Mann delivered emails to the inquiry. From that evidence and his testimony they concluded that he deleted no emails. This does not compute.

Three themes are emerging from the newly released emails:

1. prominent scientists central to the global warming debate are taking measures to conceal rather than disseminate underlying data and discussions;
2. these scientists view global warming as a political "cause" rather than a balanced scientific inquiry, and;
3. many of these scientists frankly admit to each other that much of the science is weak and dependent on deliberate manipulation of facts and data.

The Global Warming Scam

The most revealing section of Climategate was a large file with the report of a scientist called "Harry" who had the job of sorting out the files. The file is full of computer jargon, but here are some of the more juicy bits:

> *Sometimes life is just too hard. It's after midnight – again. And I'm doing all this over VNC in 256 colours, which hurts. Anyway, the above line counts. I don't know which is the more worrying – the fact that adding the CLIMAT updates lost us 1251 lines from tmax but gained us 1448 for tmin, or that the BOM additions added sod all. And yes – I've checked, the int2 and int3 databases are IDENTICAL. Aaaarrgghhhhh.*
>
> *I guess.. I am going to need one of those programs I wrote to sync the tmin and tmax databases, aren't I?*
>
> *Actually, it's worse than that. The CLIMAT merges for TMN and TMX look very similar:*
>
> *OK, this is getting SILLY. Now the BOM and CLIMAT conversions are in sync, and the original databases are in synch, yet the processing creates massive divergence!!*
>
> *OH F*** THIS. It's Sunday evening, I've worked all weekend, and just when I thought it was done I'm hitting yet another problem that's based on the hopeless state of our databases. There is no uniform data integrity, it's just a catalogue of issues that continues to grow as they're found.*
>
> *25. Wahey! It's halfway through April and I'm still working on it. This surely is the worst*

project I've ever attempted. Eeeek. I think the main problem is the rather nebulous concept of the automatic updater. If I hadn't had to write it to add the 1991-2006 temperature file to the 'main' one, it would probably have been a lot simpler. But that one operation has proved so costly in terms of time, etc that the program has had to bend over backwards to accommodate it. So yes, in retrospect it was not a brilliant idea to try and kill two birds with one stone — I should have realised that one of the birds was actually a pterodactyl with a temper problem.

I am very sorry to report that the rest of the databases seem to be in nearly as poor a state as Australia was. There are hundreds if not thousands of pairs of dummy stations, one with no WMO and one with, usually overlapping and with the same station name and very similar coordinates. I know it could be old and new stations, but why such large overlaps if that's the case? Aarrggghhh! There truly is no end in sight.:[292]

The Climategate scandal led to three official investigations which whitewashed all the scammers. Andrew Montford published an excellent summary of this continuing scandal.[293] Here is an extract:

---

[292] *Harry Report*, http://www.godlikeproductions.com/forum1/message934857/pg1
[293] Andrew Montford, *The Climategate Enquiries*, http://www.thegwpf.org/images/stories/gwpf-reports/Climategate-Inquiries.pdf (12

The Global Warming Scam

Foreword

When in November 2009 a large archive of emails and files from the Climatic Research Unit at the University of East Anglia appeared on the internet a number of serious allegations were made including:

that scientists at the CRU had failed to give a full and fair view to policymakers and the IPCC of all the evidence available to them.

that they deliberately obstructed access to data and methods to those taking different viewpoints from themselves;

that they failed to comply with FOI requirements;

that they sought to influence the review panels of journals in order to prevent rival scientific evidence from being published.

Even if only some of these accusations were substantiated the consequences or the credibility of climate change science would be immense. This was at a time when the international negotiations on climate change were foundering though not to the extent that they have done subsequently), and when, in the recession, the public and businesses were beginning to question the costs they were being asked to bear in order to achieve fundamental changes in our society. One would therefore have expected the relevant "authorities", Government/Parliament, the University of East Anglia (UAE) and the Royal

---

http://blogs.telegraph.co.uk/news/jamesdelingpole/100119087/uh-oh-global-warming-loons-here-comes-climategate-ii/

Society, to have moved fast and decisively to get to the bottom of the matter.

There was indeed a flurry of activity and three inquiries were set in train, including a hearing by the House of Commons Science and Technology Committee; the Climate Change emails Review (CCE) set up by UAE and chaired by Sir Muir Russell; and the Scientific Assessment Panel (SAP) set up by UAE in consultation with the Royal Society and chaired by Lord Oxburgh.

Sadly, as the report by Andrew Montford clearly reveals, all three reports have serious flaws. His report shows that these enquiries were hurried, the terms of reference were unclear, insufficient care was taken with the choice of panel members to ensure balance and independence and insufficient care was taken to ensure the process was independent of those being investigated, e.g. the Royal Society allowed CRU to suggest the papers it should read.

Sir Muir Russell failed to attend the session with the CRU's Director Professor Jones and only four of fourteen members of the Science and Technology Select Committee attended the crucial final meeting to sign off their report.

But above all, Andrew Montford's report brings out the disparity between the treatment of the incumbents and the critics. The former appear to have been treated with kid gloves and their explanations readily accepted without serious challenge. The latter have been disparaged and denied adequate opportunity to put their case. The CCE report stated that holding public hearings *would be unlikely to add significant value*, thereby assuming that critics would not be able to provide any additional information that would help assess the validity of CRU submissions.

This failure to accord critics rights of audience was despite the fact that Lord Lawson wrote to Sir Muir Russell when the review was first announced—specifically urging that his panel should take evidence from those outside CRU who may have been wronged.

The result has been that the three investigations have failed to achieve their objective, i.e. early and conclusive closure and restoration of confidence. The reports have been more Widgery than Saville. Writing in an article The Atlantic, Clive Crook of the Financial Times referred to *an ethos of suffocating groupthink*.

Montford makes a number of recommendations, all of which have been comprehensively ignored. The founders of the climate change scam are more powerful than ever while dissent and disagreement are more comprehensively suppressed. And, the climate change scam is firmly established in the public education system, in the Universities, the print and television media and in the political opinion polls. Much of the comment has been suppressed even from the Internet browsers.

Climategate II

On 22 November 2011, a second set of approximately 5,000 emails was released and they are included in reference 2, but the searchable index which is promised has been suppressed.[294] [295] They also are summarised links to a number of subsidiary websites and discussed by Watts.[296] Bell[297] and Delingpole[298] have

---

[294] *Tallbloke's Workshop*,
 https://tallbloke.wordpress.com/2011/11/22/breaking-news-foia-2011-has-arrived/
[295]
 http://wattsupwiththat.com/2011/11/22/climategate-2-0-the-sequel-theyre-baaaack/

[296] Watts up with That,
 http://wattsupwiththat.com/2011/11/22/climategate-2-0/

published further critical reviews. They provide further confirmation of what we know already, so I will leave with the following typical quote:

> Saturday #rd December Email 3555
> Mann: We actually eliminate records with negative correlations;
>
> Briffa: I too have expressed my concern to Phil (and Ray) over the logic that you leave all series you want in but just weight them according to some (sometimes low) correlation

---

http://wattsupwiththat.com/2011/11/30/climategate-2-0-emails-thread-2/ http://wattsupwiththat.com/2011/11/23/mr-david-palmer-explains-the-problem/
http://wattsupwiththat.com/2011/11/22/cracking-the-remaining-foia2011-all-7z-file/
[297] Bell, Larry,
http://www.forbes.com/sites/larrybell/2011/11/29/climategate-ii-more-smoking-guns-from-the-global-warming-establishment/2/11
[298] Delingpole, James,
http://www.wsj.com/articles/SB10001424052970204452104577059830626002226
http://blogs.telegraph.co.uk/news/jamesdelingpole/100119087/uh-oh-global-warming-loons-here-comes-climategate-ii/

CHAPTER 10: THE TWILIGHT OF THE GODS

THE ENVIRONMENTALIST RELIGIOUS dogma that *humans are destroying the earth* has spawned many scams. Its most ambitious project, veritably a Superscam, has been the claim that the climate is controlled by human emissions of so-called greenhouse gases. These cause global warming which will ultimately destroy us unless we cease using fossil fuels.

The Intergovernmental Panel on Climate Change (IPCC) was set up in 1988 in order to supply scientific evidence to support this scam.

It was realised from the start that the task was impossible. The earth does not have a temperature and there is no way that a scientifically acceptable average temperature can currently be derived. It is not possible to know whether the earth is warming or cooling.

Then, the climate is constantly changing. No part is ever in equilibrium. The trace gases in the atmosphere are not well mixed and their concentrations change constantly in every place. It is not possible to derive an average concentration for any of them.

The genuine science of the study of the climate, built up over many centuries as the discipline of meteorology, has officially established weather forecasting services in most countries.

These services now measure many climate properties with a variety of instruments, including satellites. The measurements are used in the most up to date computer models based on currently accepted physics, thermodynamics and statistics, adjusted for local conditions. They provide the only scientifically valid daily forecasts of future weather for every part of the earth. Atmospheric carbon dioxide has not proved to be useful and they do not even bother to measure it.

It is simply not possible to overcome these difficulties with honest science. It has therefore been necessary to employ fraud, dishonesty, distortion fabrication, massive public relations and enormous sums of money in order to claim that they have solved them.

Jim Hansen of the Goddard Institute of Space Studies, New York, provided a pseudo global temperature technique that has proved useful to the scammers. He admits that there is no such thing as an absolute Surface Air Temperature (SAT) which he calls *elusive*.

Meteorologists know it is impossible to measure a plausible average surface air temperature. Instead, they record the daily maximum and minimum in a protected screen at their weather stations. Today they often also measure at different intervals as well. These are a useful guide to temperature conditions and are plotted in their weather maps but they are not capable of providing a scientifically acceptable average.

Hansen and Lebedeff 1987 ignored what Hansen had already said was impossible. They assigned a constant temperature to each weather station for a whole month and assumed that this temperature applies also to a radius of 1200km around each weather station.

The Global Warming Scam

The chosen temperature was the total average maximum and minimum temperatures measured at that station for each month, the sum of the statistically unacceptable maximum/minimum averages.

They considered that could correlate each station figure with the next weather station. But their correlation coefficient was only 0.5 or lower. By subtracting the average from stations in all latitude/longitude boxes from the average in each box they got an annual global temperature anomaly record. There is no mention of the very large inaccuracy figures that should accompany this exercise, or of the varying number and quality of the global weather stations, both currently and over time.

The IPCC has used the supposed *trend* of a measly few decimals of a degree of this concoction to prove that global warming is happening and will inevitably rise dangerously.

Now it has broken down. This trend has hardly changed for 18 years while greenhouse gases have supposedly increased. The IPCC has resorted to desperate measures. Instead of annual warming we now have to worry about decadal warming, Efforts are escalated to fudge the figures and publicise a slight rise of hundredths of a degree as evidence of permanent warming.

The required treatment of atmospheric carbon dioxide was made by Charles Keeling of the Scripps Institute of Oceanography La Jolla California. The grossly oversimplified climate models demand that atmospheric carbon dioxide is globally constant, only increasing from more human emissions.

This was a problem because there exists some 40,000 previous measurements going back to the early 19th century, published in famous peer reviewed journals, sometimes by Nobel Prize-winners. These measurements showed that surface concentrations of carbon dioxide in the atmosphere are never constant and vary from one place to another, time of day, season, and wind direction.

Keeling suppressed this early information. He gave the excuse that he had a slightly different measurement method and he had discovered that there was a *background concentration* which was almost constant and increased steadily with increased emissions.

Keeling based his figures on sites at the Mauna Loa volcano on the island of Oahu, Hawaii and a site in Antarctica. In order to come close to a globally constant value it was required that most other measurements were made from coastal sites on winds from the ocean where any figures that did not comply with the supposed background are rejected as noise.

A difficulty was that the steadily increasing figures over the years did not easily agree with the rather sporadic behaviour of the approved global temperature record.

Now, this carbon dioxide scam has broken down. The NASA satellite AIRS system now provides frequent global maps of carbon dioxide concentration showing that it is not well mixed, is highly variable, and tends to be higher in regions of high emissions. The officially sponsored *background* is no longer relevant; the fact that the supposed warming effect of carbon dioxide is *logarithmic* with concentration means that increases have little effect in high concentration areas and are most effective over forests and pastures where they are beneficial.

The IPCC climate models defy all of the accumulated knowledge of climate science currently practised by meteorologists and replace it with a system of absurdities which has been amazingly successful in fooling the politicians and the general public.

Instead of the ever-changing climate we know, it is now assumed to be static where all heat exchanges are by radiation. Admittedly the input and output are radiation but everything else in the climate combines all methods of heat exchange, predominantly conduction, convection and latent heat change.

The Global Warming Scam

The sun is assumed to shine all day and might with equal intensity. The earth is flat and dead where living creatures are impossible except they emit greenhouse gases

All the past climate effects known to meteorology are *parameterized* and assumed to be constant.

There is no hope that such a model could possibly forecast future climate and the IPCC even admits this. They say the models provide projections, never predictions. At the beginning they avoided being proved wrong by *projecting* only so far ahead that they could be sure nobody living would survive to check.

The IPCC has now been running for 25 years and the early reports had to show that the models fitted their temperature record. Now it doesn't. Also the models could be used to calculate present upper troposphere temperatures, and that does not work either.

They are therefore in deep trouble. All they can do is prevent people from telling the truth. Every news bulletin, every newspaper must have a daily reference to global warming or carbon footprint or endure protests from climate activists who must all write letters to the press and organise rent-a-crowd gatherings of environmental devotees to picket any discussion venues There must be constant lectures by those most financially dependent on the scam. They hunt in packs to ensure everybody toes the line, maintains the consensus.

With luck, the downfall of Valhalla will take place at the Paris Climate meeting in December 2015 where the attempts to impose a global climate dictatorship will either fail miserably or fizzle slowly.

What a relief!

Vincent Gray

ABOUT THE AUTHOR

VINCENT GRAY WAS born in London and educated at Latymer Upper School Hammersmith and Emmanuel College, Cambridge, UK where he was a Major Scholar, obtained a First Class Honours Degree in Physics Chemistry Crystallography and Mathematics and a PhD in Physical Chemistry.

He has followed a scientific research career in Paris (Institut Pasteur), Canada (National Research Council), UK (Plastics, Coal, Timber, Paint, Adhesives, Building), New Zealand (Building Research, as Director, Forensic Science, Coal) and China; (Zhejiang and Kunming Universities).

He has been a Climate Specialist and Consultant for the past 23 years. He has been a Reviewer of all the Reports of the Intergovernmental Panel on Climate Change and has written two books along with many papers and articles on the subject.

He lives in Wellington, New Zealand.